T0308953

WITH A CAMERA IN MY HANDS

Oral Biography Series

Edited by William Schneider

Books in the Oral Biography Series focus on individuals whose life experiences and personal accomplishments provide an intimate view of the events, personalities, and influences that have shaped Alaska history. Each book is created through a collaborative process between the narrator, an editor, and often other community members, who record the oral history, then transcribe and edit it in written form. The result is a historical record that combines the unique art of storytelling with literary technique and supporting visual and archival materials.

volume one
The Life I've Been Living
Moses Cruikshank
Recorded and compiled by William Schneider

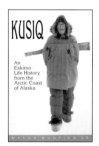

volume two
Kusiq: An Eskimo Life History from the Arctic Coast of Alaska
Waldo Bodfish, Sr.
Recorded, compiled, and edited by William Schneider in collaboration with
Leona Kisautaq Okakok and James Mumugana Nageak

volume three
With a Camera in My Hands: William O. Field, Pioneer Glaciologist
William O. Field
Recorded and edited by C Suzanne Brown

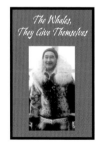

volume four
The Whales, They Give Themselves: Conversations with Harry Brower, Sr.
Recorded and edited by Karen Brewster

WITH A CAMERA IN MY HANDS

William O. Field, Pioneer Glaciologist

A LIFE HISTORY AS TOLD TO C SUZANNE BROWN

UNIVERSITY OF ALASKA PRESS —— FAIRBANKS

© 2004 University of Alaska Press

P.O. Box 756240-UAF
Fairbanks, AK 99775-6240
toll free in U.S.: 888.252.6657
phone: 907.474.5831
fypress@uaf.edu
www.uaf.edu/uapress

Printed in the United States of America.
This publication was printed on paper that meets the minimum requirements for
ANSI/NISO Z39.48-1992 (Permanence of Paper).

Paperback ISBN: 1-889963-47-X
Cloth ISBN: 1-889963-46-1

Library of Congress Cataloguing-in-Publication Data:
    Field, William O. (William Osgood), 1904-
    With a camera in my hands : William O. Field, pioneer glaciologist : a life
        history as told to C Suzanne Brown.
    p. cm. -- (Oral biography series ; v. 3)
    Includes bibliographical references (p.   ).
    ISBN 1-889963-46-1 (cloth : alk. paper)
    ISBN 1-889963-47-X (pbk. : alk. paper)
    1. Field, William O. (William Osgood), 1904- 2. Glaciologists--United
    States--Biography. 3. Glaciology--United States. I. Brown, C Suzanne.
    II. Title. III. Oral biography series ; no. 3.
    GB2403.2.F54 A3 2003
    551.31'092--dc22

                                 2003016064

Cover design by Dixon J. Jones and Mike Kirk.

Endsheets: Field Glacier on the western side of the Juneau Icefield, Coast Mountains,
        southeast Alaska (photo by Austin Post, USGS, September 12, 1986).
Front cover: see figure 254.
Back cover: see figure 198.
Front flap: see figure 233.

CONTENTS

ILLUSTRATIONS

Figures

Comparative Glacier Photographs

Historical Photographs

Maps

FOREWORD

William O. Field's life and career provide an umbrella to modern twentieth century glaciology. He was there at its beginnings in the 1920s. He has been its premier archivist into the 1990s. When not pursuing his own glacial investigations, he encouraged, coordinated, organized, counseled, and befriended several generations of alpine and polar glacier scientists. If scholarly disciplines can have founding fathers, surely Bill Field is one of the few, key elders who sired the field of glaciology.

It is not unlike William Field that in his own oral history he chose to ignore his place in the geosciences, with no mention of his host of accomplishments and contributions. The man's modesty was a trademark. Most young scientists and field workers probably are unaware of much of his work, the majority of which has remained outside the mainstream journals concerned with ice physics, mass balance, and glacier climatology. But those whose own career development coincided with the adolescence and maturing of glaciology know his publications and recognize the ways in which Field gave the fledgling field direction, organization, access to funding, and, indeed, a place literally to call home.

We are fortunate to have this oral history. Suzanne Brown is to be congratulated for extracting, from a surely often reluctant subject, such a trove of reminiscences and biographical information. Bill Field had his contradictions: we knew him to be consistently warm and outgoing, yet he was also a private man and even those who had been close to him for decades will discover much that is new here. My own first encounter with Bill Field was as a freshman down to The City from New Haven, wandering into the American Geographical Society in search of mountaineering literature. Luckily, the first person I encountered was Bill Field and, unknowingly at the time, was led into the world of snow and ice science. Bill Field's friendship and mentorship was a constant in the succeeding four and one-half decades. It is a privilege to write this foreword and introduction for his oral biography.

Mel Marcus
Tempe, Arizona

PREFACE

The Man

On a recent fishing trip with my family to the Kenai Peninsula in Alaska, I couldn't believe how much I felt the presence of my friend and colleague, Bill Field. As we passed Spencer and Bartlett glaciers at a distance on our way to Homer, I found myself explaining that these were the first glaciers in Alaska Bill saw up close, in 1925, both glaciers then ending only a few hundred yards from the railroad. The railroad is actually built on the 1885 moraine of the Spencer Glacier, a fact that gave Bill pause to wonder what will happen if the glacier advances again.

At Trail Glacier, I recounted how Bill and his two friends camped in a mosquito-infested alder patch for three days and climbed a small peak nearby, with the "grand thought" that perhaps they were the first to be there. As we continued along the Sterling Highway, it was like following Bill as he journeyed by boat that year down Kenai Lake and then the Kenai River, and across Skilak Lake with his guides George Nelson and Hank Lucas to their hunting cabin on the southeast side of the lake. When we passed Tustumena Lake, I remembered hearing about Bill's first airplane flight in 1927, going from Anchorage to Tustumena Lake in a plane called Alaska #1, flown by Russell Merrill for whom Merrill Field in Anchorage is named. Passing Kasilov made me laugh as I remembered Bill's humorous account of how he had taken a boat, the only means of transportation then, from there back to Anchorage at the end of that 1927 trip, and between the rough water and the smell of diesel fumes inside the boat, he thought he wasn't going to make it back alive.

It then occurred to me that I am reminded of Bill Field when I travel anywhere in Alaska. When I look at maps of Alaska, I wish I knew the names of the villages, glaciers and mountains half as well as he knew them. Of course, he named many of the glaciers himself. And the people he has worked with: Andy Taylor, mountaineer; Percy Pond, noted photographer; Lawrence Martin and H.F. Reid, pioneer scientists who worked in Alaska. To many of us, these people are legends; to Bill, they were friends. There are but few who can claim personal relationships with the pioneers in their particular field. Bill was a living link to the past, to these pioneers and their early work in glaciology, and Bill's personal knowledge of these people, of the context of their research, and his exposure to them enabled him to take full advantage of their experiences.

I knew Bill for eighteen years. I am a glaciologist by training with an interest in the history of glacier observations in Alaska. I met Bill through my friend Austin Post, a legendary figure in aerial glacier photography. Bill stopped in Seattle on his way home from Alaska in the summer of 1976 and we had dinner together. I remember in our conversation at that dinner Bill's comment about needing help organizing his collection. When I offered my assistance, little did I know it would involve almost twenty trips across the country from Tacoma, Washington, to Massachusetts; several field trips to Alaska; the preliminary archiving of Bill's vast collection of photographs, manuscripts, field notes, books and maps; and, probably most challenging, the recording, editing and publication of this oral biography. Over the years I have asked Bill many questions and have eagerly listened to his stories. This book is an opportunity to share with others some of the experiences that Bill shared with me. I have never actually considered any of this to be work, but more of a privilege—to know and work with the man and his collection, to listen to his stories, and edit this book to share Bill's life with others.

The Series

This book is one in a series of oral biographies published by the University of Alaska Press. The subjects of these books are individuals whose life experiences and personal accomplishments provide a glimpse, first-hand, into the events and personalities that have shaped Alaska history. An oral biography is a person's story in his or her own words. The oral recordings are transcribed, compiled, and then edited into a book that is designed to read as much as possible the way the narrator talks. This oral biography has several appendices that contain information valuable to the reader that either was not included in the recordings or removed from the initial text for the purpose of space and flow of the story. Anyone desiring even more of Bill's story is encouraged to review the transcripts of the original tapes.

I was asked to edit Bill's oral biography by Will Schneider of the Oral History Department at the University of Alaska Fairbanks. He thought it would complement Bill's collection, which is now housed in the archives of the Alaska and Polar

Regions Department in the Elmer E. Rasmuson Library at the university. Bill's collection was one of the two largest collections ever received by the archives. Besides the fact that Bill became a world-renowned scientist as well as the undisputed father of glaciology in North America, reading Bill's life story in conjunction with using his collection helps create a better understanding of exploration in Alaska in the late 1800s and the first third of this century, of the people involved, and of the hardships they faced.

One cannot truly appreciate the present without knowing about the past. Bill was a keen observer and photographer. His knowledge and memory of people and events integral to the history of exploration in Alaska was almost unsurpassed. In many ways Bill's life is a representation of the history of northern science, beginning with the explorer/scientist, evolving into a focused institutionalized science, and ending up as big business science.

I am neither an oral historian nor a professional writer. I agreed to edit this story only because I knew and worked with Bill, I was familiar with all his stories and the people involved, and I know the extent and content of his collection. As I began to work on the project, I realized that this would not be a simple, one-track life story. This fact has created tremendous problems in compiling and editing into a readable fashion his life and accomplishments, illustrating the many dimensions of Bill Field and how they contributed to the personality who will be remembered as the father of present-day glaciology in North America—the man who encouraged scores of young people to study glaciers. Naturally in this book there is an emphasis on his glaciological work, but much of the rest of his story also is included so people will have a better understanding of Bill Field the man and thus a better appreciation of Bill Field the scientist.

As well as being an accomplished mountaineer, photographer/filmmaker, and natural scientist, he was knowledgeable in world history and deeply concerned about foreign affairs and cultural relations. He was concerned, too, with attempts to achieve understanding between people of other nations on a strictly nonpolitical basis. Perhaps most of all, however, he was keenly interested in *documenting life—and thus change—with a camera.* He literally went through life with a camera in his hands.

Bill's life has been one rich in experiences: paddling a canoe to the glaciers while today we hop into a helicopter; traveling by pack train in the Canadian Rockies where now one speeds along a major highway; sharing a tent in Alaska with well-known mountaineer Andy Taylor for three weeks; spending a few days with the famous

Alaska photographer Percy Pond at his cabin near Taku Glacier. Of course, Bill's travels weren't just to Alaska. He worked in the film industry for a number of years, traveling to Africa, Cuba, and Italy. He made extensive trips of his own to the "hidden valleys of the Caucasus" in 1929 and 1930. He went there initially to see the mountains, but quickly discovered the fascination of the area as an ethnological museum, with a society emerging from the twelfth into the twentieth century. With his increasing interest in documenting life with photographs, he took many pictures on his two trips to the Caucasus in 1929, and many are included in this book.

This man was a font of living history, whether one's interest was early Alaska, glaciers, the Canadian Rockies, filmmaking, the Caucasus, foreign affairs or global climate change. In reflecting on the many dimensions of Bill Field, I have wondered if there is a place now for careers that integrate very different interests yet in the long run contribute much to what we recognize as a specific field of inquiry. Could in fact a Bill Field emerge in this day and age?

How This Book Is Organized

I have edited and arranged Bill's life story as follows. Chapters 1 and 2 describe the formative events in his youth and set the stage for understanding other dimensions of his life. They trace Bill's path as he begins to pursue some of his developing interests: excursions and first ascents in the Canadian Rockies; climbing in the Alps; his first trip to Alaska, followed by his research at the Harvard Library on the history of glacier observations in Alaska. By the age of twenty Bill was an accomplished explorer, mountaineer, and photographer. The photographs at the end of the first chapter show the Canadian Rockies as they were seventy years ago, when roads were almost unknown, pack trains were the mode of transportation, and first ascents were still plentiful.

Chapters 3 (the 1926 trip to Glacier Bay) and 6 (the 1931 trip to Prince William Sound and 1935 trip to both Glacier Bay and Prince William Sound) chronicle how Bill started on his distinguished, life-long career in glacier monitoring. For most of his early scientific trips, I have tried to include not so much what was done, but how it was done and the circumstances surrounding the events. As is typical of a person's memory of first-time events, Bill's detailed description of those trips was fantastic. In addition, when Bill described those trips sixty years later for this oral biography, he was looking at them in relation to what developed later in his life, and thus their significance to his later life and career crystallized. Chapter 4 describes Bill's last family trip to Alaska in 1927. It is included for those who know the significance of a flight with Russ Merrill, and for those Appalachian Mountain Club members who visited the Pinkam Notch hut in the 1930s and came to know and love the resident dog Hiu Skookum.

In 1927 the opportunities for work for Bill were not in glaciology. He was not ready to embark on a career in glaciology at that time anyway. His 1931 trip to Prince William Sound and his 1935 trip to both Glacier Bay and Prince William Sound as well as his working with François Matthes on the Committee on Glaciers during those years were fundamental to Bill's gradual professional evolution, culminating in a job with the American Geographical Society in 1940. Thus, during the period 1927 to 1935 Bill had a job involving world travel and filmmaking—also two passionate interests, as Chapter 5 details. Even though this chapter interrupts the flow of the story of Bill's development as a glaciologist, it was necessary to include because first, his progression from neophyte to professional was not continuous, and second, the trips he made during that time were very important to Bill, especially his two trips in 1929 to the Caucasus. The bulk of Chapter 5 is a selection of rare photographs taken by Bill on those trips. He always recounted his experiences on these trips by paging through his Caucasus photo albums and I thought it fitting to do the same here with photos selected from those albums.

Bill's years at the American Geographical Society (AGS) are described in Chapters 7, 8, and 9. Bill wanted to live a productive life, and when the door opened a little into the AGS, with a "desk in an alcove," he saw an opportunity to do something useful and he made the most of it. Shortly after he joined the AGS, his career took on an international perspective, as evidenced by his adoption of the metric system of measure. Bill's story ends with his reflections on his career, with the last page a copy of the memorial written by Greg Streveler of Glacier Bay as only Greg can write.

A few quotations appear throughout the book; most of them are ones that Bill collected but a few are quotes from his friends. Anyone involved with glacier monitoring will appreciate the significance of each of them.

Bill told stories with photographs, and thus I likewise have used many photographs to tell *his* story. In addition, as a special tribute to Bill and his work, two sections of photographs from selected from Bill's collection is included. The Comparative Glacier Photographs are some of the most spectacular examples of the scientific contributions of Bill's photographic methodology, and the Historical Photographs demonstrate the additional human and historical value of the archive.

It is a well-known fact that Bill was a very modest man, and many readers will understand and appreciate the difficulty I encountered in trying to encourage him to talk about his accomplishments in the field of glaciology. When I once asked him why he always used "we" when describing his field trips, even when he was by himself, he replied, "It was always a team effort, not just my work. I am tired of people using 'I' all the time." In view of this graciousness and selflessness, I asked Dr. Mel Marcus, Professor of Geography and Environmental Studies at Arizona State University, if he would write the introduction for this book, explaining the significance of Bill's contributions to and accomplishments in glaciology, as well as Bill's place in the history of modern glaciology. Mel's friendship with Bill and Mary spanned more than forty years, and he attributes his scientific beginnings to Bill's guidance and teaching. Thus, Mel's piece is written from a personal perspective as well as from that of a scientific colleague, and it is a marvelous tribute to Bill and his work.

Glaciologists went to Bill Field, his collection and his memory to establish a reference for their research. Any book concerning him should also be a reference. To that end I have tried to include more than just cursory information in the notes and appendices in hopes that this book will serve as a small resource to the research community, as Bill would have wanted. Appendix C outlines every scientific trip Bill made to Alaska and the Canadian Rockies, and includes the names of most members of the party, vessels chartered and stations occupied (based on information available). For young glaciologists making their first trip into the field, perhaps to retrace some of Bill's footsteps, I hope this book will give them pause to think about how far the study of glaciers has come, and to consider the contributions of Bill Field to the creation and preservation of a long-term record of glacier monitoring, an effort that continued to the day he died.

After Bill's trips to Glacier Bay in 1926 and Prince William Sound in 1931, he discovered there was no central collection of glacier photographs taken by people on earlier expeditions. So Bill saw the need to assemble in one place as many of these photographs as possible for his projects as well as "for the general interests of glaciologists in this country." He began this project in 1932 and kept adding to it for the rest of his life. In the 1980s Bill began to look for a place where his work would be preserved yet accessible and, he hoped, extended. He was deeply concerned that the work not be lost and that it continue to be available as a resource for anyone in need of information. He eventually chose the archives of the Alaska and Polar Regions Department at the Rasmuson

Library of the University of Alaska Fairbanks because of its commitment to research in the polar regions as well as its state-of-the-art facility. A detailed description of the collection is given in note 13 of Chapter 9. As a last act of generosity, he helped assure that his collection was preserved and made available to others by partially funding the processing of his collection. That processing is underway now and the most useful parts of the collection will be available shortly.

Bill passed away on June 16, 1994. Some time before his death, I offered to push the book ahead and have it published while he could still enjoy it. He told me no—to take my time and do the best job I could. Well, Bill, the project has taken a little longer than I anticipated but I hope you like the book. It has been both an honor and a pleasure to work on it.

Photo Numbering System

Bill took the majority of photographs in this book and all but a few bear his unique photo identification number, such as F-31-296. The letter on the left indicates who took the picture; F in this case stands for Bill Field. There are a few instances of WF, which stands for Bill's father, William Field. Likewise, on some of Bill's AGS field trips, other members of the party took pictures for Bill and their initials appear in place of the F. The middle number is the year in which the photograph was taken. In the example above, that would be 1931. The number on the right is the number of the exposure taken that year. In the example, the photograph was the 296th taken in 1931. If there is more than one exposure number separated by a comma, it means the figure is composed of two or more photos. Occasionally a letter precedes the exposure number. This was sometimes used to indicate that the photos were taken on a special trip. For example, the C in the photo numbers in Chapter 6 stands for the Caucasus. Some of Bill's more recent photos are labeled with a K, which indicates the photo was taken with the camera Bill used for color slides (Kodachrome film); or an R, which means Bill used the comparatively lightweight Rolleiflex camera (black-and-white film).

Acknowledgments and Dedications

Many wonderful people have helped me over the years with this project; people generous with their time, information, encouragement, and support. Without their help this project would never have been finished and I wish to thank them profusely.

I needed assistance mostly in preparing the many photos and maps for inclusion in the book. David Hirst of the United States Geological Survey (USGS) has been available throughout the whole process for my many questions and to help when possible with photo preparation. He was always there with moral support when I needed it, too. I also owe many thanks to Carolyn Driedger, my friend, former colleague and confidant for many years, and Dave Wieprecht of the USGS for their help with photo and map reproduction. I don't know what I would have done without you both! In addition Dorothy Hall at NASA assisted in obtaining reproductions of some of the larger photos. A few of Bill's friends offered photographs of their own for use in the book. I thank Ed LaChapelle for his pictures from the Juneau Icefield in the 1950s, Mark Noble for a great picture of Bill in Glacier Bay, and Carl Benson for his picture included near the end of the book. A big thank-you also goes to Mark Meier, who provided me with several key photos of European scientists on very short notice. A large number of family photos are also included and I wish to express deep gratitude to Bill's children John and Diana Field for allowing me to use those photos.

Obtaining the information for the notes in a book such as this are always a problem and I am indebted to Carol Edwards at the USGS library in Golden, Colorado, for her assistance in obtaining biographical information of the USGS people from long ago that Bill mentions. India Sparks at the Alaska State Library in Juneau provided much assistance in researching non-USGS people. I wish to thank Ted Hart, Director of the Whyte Museum of the Canadian Rockies, for providing information on the Canadian Rockies and for being such a pleasant host during our visits to the museum. I always enjoy stopping in and look forward to my next visit, I hope in the near future.

Map preparation can also be difficult, especially to the severely artistically challenged like myself, and I profusely thank Dee Molenaar who, even with a very busy schedule, graciously agreed to draw some of the maps and even taught me a little technique. In addition, Dee gave me the intensive editorial review I was looking for. While reading the manuscript prior to preparing the maps his pencil could not sit still—I had forgotten that he was a technical editor for the USGS for twelve years! Dee has continued actively to support me in the completion of this project and without his encouragement I truly believe this book would not exist.

I shared the manuscript with my father, Charles S. Cummings, and thank him for picking up a few inconsistencies and points of grammar. Carl Benson's review led to the inclusion of the magnificent piece by Mel Marcus and to a reshaping of the whole book. Your input, Carl, was a turning point in the preparation of the manuscript!

When I first began to travel to Massachusetts to help Bill organize his collection, I was working in the Geological Survey's Project Office Glaciology. I gratefully acknowledge the generous support of my efforts by my boss Mark Meier, as well as his interest in the progress of this book during our talks on my trips to Boulder, Colorado, in recent years. I always look forward to spending time with Mark and his wife Barbara during these visits, and always learn something new.

I felt very privileged when Mel Marcus agreed to write the foreword and introduction for this book and was deeply saddened by his untimely death in 1997. He was a special friend of mine. I am also fortunate that another special friend of mine, Greg Streveler, allowed me to include his beautiful memorial to Bill as the ending piece for the book.

While editing this book, it has been a comfort to know that Bill's son John supported my efforts and I gratefully acknowledge that and thank him for including me on recent trips to Glacier Bay to reoccupy his father's photo stations. Those have been great times that I will always cherish. This book has passed through the hands of several editors over the years. I especially wish to thank Jennifer Collier and Erica Hill at the University of Alaska Press for the outstanding job they have done. It has been a sincere pleasure to work with them. I am also grateful to William Schneider at the university for his support throughout the preparation of the manuscript. Almost last but not least, I must express extreme gratitude for my husband Tom, who has supported us and allowed me to donate so much of my time, energy, and resources to this project.

Finally, I wish to dedicate this book to four very special people:

Austin Post, who first introduced me to Bill Field.

Captain Jim Luthy, skipper of the National Park Service vessel *Nunatak*, who has worked tirelessly to keep Bill's memory and legacy alive in Glacier Bay.

Ruth Ide, who selflessly helped Bill for many years, and helped me over the last fifteen as I began to process his collection as well as edit this book.

Ian, my son, because this is the first and will be the last book your mother ever writes.

C Suzanne Brown
May 2003

You know, Jim, life has been a success when you have a happy memory.

—Conrad Kain to Jim Simpson, Sr.,
Bow Lake, Alberta

INTRODUCTION

The Early Days

The serious study of glaciers is a relatively new business. From Carpentier to Agassiz with his "Glacier Theory," only a few natural scientists, mostly European, turned their attention to glaciers or the landscape-forming work they accomplished. By the late nineteenth century, however, American scientists were beginning to understand the immense erosional and depositional handiwork of the great continental glacier system. Soon the rudimentary patterns of glacial action had been roughly sketched across New England, the midwestern American prairies and Canadian plains, and eastern Canada.

If the world of continental glaciation was falling into focus, the observation and study of smaller mountain and piedmont glaciers lagged behind. Also, the early research focused on the work accomplished by glaciers (glaciation, glacial geology) and not on the properties and processes of living glaciers (glacierization, glaciology). It was not until the spectacular tidewater and piedmont glaciers of coastal Alaska became the focus of research in the 1890s that a significant number of researchers began to look at the glaciers of North America in the present tense.

A variety of factors, converging over a two-decade period, brought (for those times) a veritable explosion of research activity to the Alaska Gulf Coast glaciers. These included relatively easy access by sea, the Klondike gold rush, the international boundary survey, interest in surveying terrain and resources in this largely unexplored American territory, and the theme of scientific exploration that no doubt appealed to expedition patrons such as W.A. Harriman and the National Geographic Society. One can also imagine the romantic drawing power of these magnificent ice-covered ranges, especially as expressed in the prose images of John Muir. Whatever the reason, the region was visited, observed, and researched by a veritable Who's Who of American geography, geology, and exploration: H.F. Reid, G.K. Gilbert, I.C. Russell, R.S. Tarr, L. Martin, W.C. Mendenhall, A.H. Brooks, the resolute surveyors of the boundary survey and others.

William O. Field's role in this chronology was critical. By World War I, most Alaska glacier research had languished. With the exception of Tarr and Martin's 1914 opus, *Alaskan Glacier Studies*, and W.S. Cooper's continuing ecological work in Glacier Bay, few scholarly treatments were published after 1910. There was a hiatus in the study of Alaska glaciers and a continuing paucity of work from the conterminous forty-eight states. With Bill Field's 1926 visit to Glacier Bay to ascertain glaciological changes that had occurred since Reid's work in the 1890s, the link between the scholarly past and future was established. In a sense, this was a renewal for American glaciology. Henceforth, there would be careful, regular monitoring of glaciers in coastal Alaska.

Thus, while modern glaciology was experiencing its birth in Scandinavia with the process-oriented investigations of Sverdrup and Ahlmann in the 1930s, it also began to emerge in North America with the glacier measurements, photography, and record keeping of William O. Field. Inevitably, the two converged because understanding and explaining glacier process on the one hand and measuring and maintaining a glacier data base on the other are the yin and yang of glaciology. These two must be folded together; field observations provide the stuff on which theory and models are constructed and tested.

The AGS Years

The geophysical and climatological counterpoint to Bill Field's continuing monitoring program truly emerged after World War II. His Department of Exploration and Field Research at the American Geographical Society (AGS) became the hub for much of the national research planning and organization in the growing glaciology discipline. It must be remembered that the ranks of qualified snow and ice scientists then were thin. In the United States, the leaders were Robert Sharp at the California Institute of Technology, Henri Bader at the U.S. Army Snow, Ice, and Permafrost Establishment (SIPRE, later CRREL), and Bill Field at the Society. There was, of course, a fairly large corps of distinguished glacial geologists attending to the business of glacier erosion and deposition.

In any event, Bill's Society office at 156th Street and Broadway in upper Manhattan became both a mecca and coordinating center for much of American glaciology. Several factors combined to attract an array of glacier scientists, alpinists, and polar researchers (amateur and professional alike) to the AGS over the next three decades: the Society's and Bill's personal photographic collections, the AGS map library, and the AGS library which was arguably the best geography-earth science library in the Western Hemisphere.

As glaciology developed in the 1950s, culminating in the International Geophysical Year (IGY) in 1957–1958, Field's office was an informal, de facto headquarters for the discipline. Indeed, if one was at all connected to glaciology, one simply did not visit the eastern seaboard without dropping in on Bill and his staff to glance at the latest library or photo acquisitions and to gossip about the world of glaciers. Few visitors left without some new insights into the business of glaciers or new professional contacts. On a good day, a lucky young researcher might find an Ahlmann of Sweden or Thorarinsson of Iceland or Perutz of Great Britain as his luncheon partner.

In a 1977 commentary in the *Geographical Review*, I suggested:

> Such communication was especially important to the emerging field of glaciology. Glaciological practitioners range from geologists, geographers, and botanists, to physicists, chemists, meteorologists, and engineers. Glaciology is multidisciplinary or interdisciplinary, and therefore it is dependent on the ability of people to communicate outside their own disciplines. The Society's function [via Bill Field] was significant in this regard, and it is remarkable that with all the interruptions so much was accomplished in the manner of serious research, publication, and data compilation.

The best known AGS research from this period included: the multifaceted Juneau Ice Field Research Project; Cal Heusser's geobotanical studies of northwestern North America, much of the field work having been shared with Bill Field in Alaska and the Canadian Rockies; the nine 1:10,000 "representative glacier" map series done for the IGY; and the organization and operation of Glaciology World Data Center A (for the Western Hemisphere). The young scientists whose work directly or indirectly could be linked to Bill included such later luminaries as Calvin Heusser, Ed LaChapelle, Maynard Miller, Larry Nielsen, Ed Thiel, Austin Post, Dick Cameron, James Case, Colin Bull, Carl Benson, and John Mercer.

The planning of many glaciological projects was initiated in Bill Field's office. As an example, Carl Benson has recounted a meeting on October 8–10, 1952, to discuss plans for research in Greenland and on the Ward Ice Shelf of Ellesmere Island. It included representatives of SIPRE (Carl Benson and Robert Schuster), the Arctic, Desert, Tropic Information Center (G. William Holmes

and Donald Shaw), the Arctic Construction and Frost Effects Laboratory (Kenneth Linnel and Edward Blackey), and the Mint Julep Project in Greenland (F. Alton Wade). Two research projects in Greenland, involving C. Benson, R. Schuster, and Ed LaChapelle as well as a research expedition with E.P. (Bill) Marshall from SIPRE and led by G. Hattersley-Smith of the Canadian Defense Research Board followed from this meeting. These were typical results. It was natural for agencies, indeed governments, to organize planning meetings in Bill's office. Yet, as time moves on, these types of substantial contributions by Bill Field tend to get lost because they do not appear in the scientific literature.

Field's guiding hand and counsel were important in the late 1950s and the 1960s. With the Arctic Institute of North America, the Society was cosponsor of the ambitious, interdisciplinary Icefield Ranges Research Project in the St. Elias Mountains. The Cartography Division and the Department for Exploration and Field Research were major players in producing the International Hydrological Decade's Antarctica Map Series. And, significantly, Bill Field was editor of the 1958 (mimeographed) *Geographic Study of Glaciation in the Northern Hemisphere*. This U.S. Army Quartermaster Corps document was the first-ever inventory of glaciers. A companion report edited by John Mercer was issued for the Southern Hemisphere in 1967.

One of Field's major accomplishments was the bringing together of the three-volume work *Mountain Glaciers of the Northern Hemisphere* (William O. Field, ed., 1975, U.S. Army Corps of Engineers, Cold Regions Research and Engineering Laboratory, Hanover, NH). The first two volumes provide a veritable encyclopedia of data, interpretation and key map and bibliographic references for each region or mountain area. Much of this wealth of detail is not available elsewhere. The third volume is a forty-nine-plate atlas of mountain glaciers. Even in the current era of remote sensing and GIS, this remains the definitive resource.

Bill Field has left another legacy for glaciological scholars: his comprehensive collection of film, maps, books, and field notes relating to the glaciers of Alaska. These materials, which arrived in November 1993, now reside in the William O. Field Collection at the Alaska and Polar Regions Department, Elmer E. Rasmuson Library, the University of Alaska Fairbanks. Scholars have been presented a rare, rich resource. Fifty-one boxes and one four-drawer file cabinet are from his film vault alone. They include a remarkable range of images—hand-colored lantern slides; motion picture film; glass plate negatives; and a full array of slides, prints, and negatives. The remaining 263 boxes include photographs by earlier observers, along with maps, books and field observations for glaciers of the Canadian Rockies and Antarctica as well as Alaska.

Honors

Bill Field garnered a number of distinguished awards over the years. These include an honorary doctorate from the University of Alaska Fairbanks, and the Busk Medal of the Royal Geographical Society. Consistent to the end, he chose not to talk about such honors in his oral history. For those who might not otherwise realize the high regard in which geoscientists hold William O. Field, some commentary is included here from two distinguished awards: the American Geographical Society's Charles P. Daly Medal and the International Glaciological Society's Seligman Crystal.

In receiving the American Geographical Society's Charles P. Daly Medal, Field joined an august fraternity. A sampling of earlier medal recipients reveals such luminaries as Robert E. Peary, G.K. Gilbert, Vilhjalmur Stefansson, Roald Amundsen, Francis Younghusband, and W.S. Cooper. The medal citation, presented at the Society's annual dinner on November 21, 1969, reads:

> For more than forty years, since before his graduation from Harvard in 1926, William O. Field has been occupied and preoccupied with the glaciers of Alaska, large and small. He has examined them, analyzed them, photographed them, mapped them, and written about them. He has studied them not only in terms of their individual behavior but in a wider context as indicators of climatic change. Perhaps no man living knows more about their regimes and characteristics or has made greater contributions to the glaciological literature of this area. In 1940 he joined the staff of the American Geographical Society and in 1948 became head of the Department of Exploration and Field Research. Since then it is probable that among the glaciological fraternity, at least, the Society's principal claim to distinction has been as the professional home of Dr. Field. Over the years his knowledge and experience have caused him to be drawn frequently into the broader realms of science. He has served on, or headed, various panels, committees, and commissions that have brought him in touch with colleagues in the geophysical disciplines throughout this

country and abroad. He played an important part, for example, in the planning of the International Geophysical Year.

> In these ways his career is a matter of record and repute. What cannot so easily be assessed is the influence he has had in generating original research on the part of young scientists who have been with him in the field, learning the techniques of precise survey and observation, and coming to share his fascination with Alaska's "rivers of ice."

> It is uniquely appropriate that the American Geographical Society should confer on Dr. Field the Charles P. Daly Medal, which is awarded "for valuable or distinguished geographical services or labors." He has both served and labored, with results that reflect the highest credit on himself and the Society.

In his response, Dr. Field, in his usual way, suggested that "many others by rights should share" with him this honor.

The Seligman Crystal is the premier honor in glaciology. The International Glaciological Society bestowed the crystal on William Osgood Field in 1983. Reporting on the award and citation, *ICE* (1983, no. 2/3, p. 46), the publication of that society stated in part:

> Many glaciologists consider William Osgood Field the integrator and catalyst of North American Glaciology.... When he returned to the AGS [after the war] Bill helped initiate and direct glaciological projects, including the Juneau Icefield Research Project and studies of the fluctuations of some 200 glaciers in Alaska, the Southern Andes, Greenland and Western Canada.

> During the IGY, he visited the U.S. bases in Antarctica. He also directed the World Data Center A for Glaciology... he was Vice President of the Commission of Snow and Ice from 1960–1963 and of the Glaciological Society from 1962–1964. He has been a staunch supporter of the Society and was made an Honorary Member in 1970. Field Glacier in the Antarctic (67°09' S, 66°23' W) is named after him.

> While his contributions are great in unraveling the story of North American glaciers, many people consider that his most valuable gift has been his service on committees and panels.

> He has been a member of the Committee on Glaciers, Section of Hydrology of the American Geophysical Union, for most of its existence since 1931 and was Chairman

from 1948–1954. He served as reporter on glaciology of the U.S. National Committee for the IGY and later became Chairman of its Technical Panel on Glaciology. He continued when that panel became the Glaciology Panel of the Committee on Polar Research, National Academy of Sciences/ National Research Council.

[I]n giving Bill this award, his colleagues have paid tribute to his pioneering work, his cataloging of glaciers, his leadership in developing world-wide programs and his positive influence on many young scientists.

The 1966 commentary in *ICE* commented that the Glaciology Panel was one of the few IGY panels still active. Bill Field's leadership helped make that possible and the glaciological community recognized that to the degree that, "Membership now rotates, but the Panel refuses to allow its original chairman to step down." *In 1995, Bill Field's Glaciology Panel is the only one of the former panels to survive and achieve committee status within the Polar Research Board of the National Academy of Science.*

The Person

When all is said and done, and for all his accomplishments, it is the person, the spirit, of Bill Field that remains with us. I am sure his host of friends could provide chapters of reminiscences: the summers in Glacier Bay; long days of conversation in the AGS offices (while his own work waited); wonderful luncheons, invariably hosted by Bill, attended by an international assortment of characters who just happened to drop by the Society; the stories of those who worked with him in coastal Alaska, the Canadian Rockies, and the Caucasus, or shared the camaraderie of international conferences and field excursions.

When discussing a basically good and kind man, it is easy to fall into the trap of maudlin praise. I know that Bill Field would have hated it. We should not forget that this was a robust man, who, while pursuing his research, dealt with the elements head-on. The coastal waters and mountains of northwestern North America are not always hospitable places. Interminable rain, rough seas, and thick brush are only a few of the impediments to effective research. And given the need to make hay while the sun shines (or doesn't) during the short, high-latitude field season, the work schedule is often exhausting. Bill Field lived and thrived in those field conditions for over six decades.

Bill Field was also a man who looked forward to a double manhattan before dinner, approached a range of international cuisine with gusto, and always had something to say on the latest political or social news. A typical evening with Bill and his wife Mary was a chance to delve into the history of film, berate some current politician, discuss literature old and new, or even relive the emergence of jazz and swing across America. Mary, a documentary film pioneer in her own right who set her own gender agenda before the term Affirmative Action existed, had a brio that kept the conversational pot bubbling. And if glaciers were absent from the conversation, their photographic and painted images took it all in from the nearby walls where they hung.

My last visit with Bill was in December of 1993. He and Mary temporarily resided in an apartment complex in Lenox, Massachusetts, where they were spared the full winter rigors of their country home and his study outside Great Barrington. He was clearly frustrated that this sojourn was keeping him from a full-bore work schedule in his office and his efforts to finish cataloging his glaciological collection. Nevertheless, nearing his ninetieth year, he was busy with an ongoing and worldwide professional correspondence and involved, as always, with map and photo interpretation. I was amazed at the range of scientists who had been in touch with him in the few months preceding our visit. It was business as usual, the glacier community seeking information and counsel from the man who had been providing it for much of this century.

William O. Field, Jr. passed away on June 16, 1994. Mary Field followed some ten months later. Active until his last weeks, Bill was informed that his most recent paper, "Changes in the Glaciers of Glacier Bay, Alaska: Using Ground and Satellite Measurements," coauthored with Dorothy Hall and Carl Benson, was to appear in a double-issue volume of *Physical Geography* as the William O. Field Festschrift. Although he did not live to see the full volume in print, he was touched to learn of this honor his friends and colleagues had planned for him.

Mel Marcus

FIGURE I

CHAPTER 1
EARLY HISTORY
1700–1924

Ancestors

My family has been in this country since the seventeenth century, primarily in New York City. My great-grandfather on my father's side, Samuel Osgood, graduated from Harvard in 1790. He knew George Washington and was the first American officer in the Revolutionary War to receive surrender from his British counterpart. He helped frame the Constitution [of the United States] and later served as the first postmaster general of the United States. Great-grandfather Bradhurst, also on my father's side, had a farm in upper Manhattan and was quite a prominent person. His farm spanned the upper heights of the 155th Street area. On the east was the low, level ground of Harlem, on the west was the Hudson River. The area around Harlem was quite swampy and there was a lot of malaria there. My great-grandfather was also a pharmacist and I heard stories about the people coming up to his farm all through the night for quinine pills.

Grandfather Field I know very little about. He died in 1888. My grandmother Augusta Curry (Bradhurst) Field I knew well, and she was very close to me in my youth (fig. 1). She was reasonably well-to-do but lived simply. She was a great person to remember historical events and she'd tell me, "Remember these things." She had known General Sherman after the Civil War and she used to say that he had told her that she would live to see the day when there would be no buffalo in the west. At that time, in the second half of the nineteenth century, there were millions of them roaming the plains.

Grandfather and Grandmother Field had two children, my father and my aunt. My aunt, Mary Pearsall Field (fig. 2) was a little older than my father. My father was named William Bradhurst Osgood Field. The summer before he was born, his father and mother were visiting in Paris. During that summer, the Franco-Prussian War broke out. When the Germans appeared over the horizon and threatened Paris, the family moved to Geneva, Switzerland and my father was born there on September 16, 1870.

FIGURE 1.
My grandmother Augusta Curry (Bradhurst) Field at Westfield Farm, Mohegan, New York, 1907. I called her "Mia," which is Italian for "my own." (WF-M-07-26; photo courtesy J.O. Field.)[1]

FIGURE 2

FIGURE 3

FIGURE 2.
*My cousin Harry Bradhurst
and my aunt Mamie Field
(Mary Pearsall Field) at
Westfield Farm, Mohegan,
New York in 1907.
(WF-M-07-34; photo
courtesy J.O. Field.)*

FIGURE 3.
*My grandfather W.D.
Sloane at Elm Court,
Lenox, Massachusetts
in 1907. (WF-L-07-L-8;
photo courtesy J.O. Field.)*

Mama and Papa

My father, Papa, or Pops as we usually called him, was brought up in part in Europe. His mother and father used to make extended visits to Rome and other parts of Europe. During those long visits, Papa went to school there. I remember his telling us how he once ran away from school in Germany and joined the circus. It was six weeks before his parents located him. He was just learning to be a bareback rider when he was yanked out and sent back to school. But in Europe he learned a great deal about European culture—the arts, and the museums and their treasures. In later years, he had an extraordinary number of intelligent, professional people for friends, primarily from the art world. He should have been one of them. He should have been a museum curator instead of an engineer. Papa was a very interesting and very intelligent man but also a difficult man, very prejudiced with strong likes and dislikes. Papa was also a collector, and when we were youngsters he had already begun to collect a library. That library was really his pride and joy. Before he died, he gave this collection to the Houghton Library at Harvard.

Papa went to Harvard for one year, and then his father died. In order to be near his mother and sister, he moved back to New York, and graduated from Stephens Institute of Technology in Hoboken, New Jersey, in 1894. He played football for Stephens Tech, a guard, and he always enjoyed telling me how they beat Princeton once. Papa was a civil engineer by profession. For some years he worked on the New York Central Railroad and traveled all over New York state making inspections. He was one of the first people to illustrate his inspection trip reports with photographs. He was a rather charming, educated man who made friends quite easily, and he met a few of the people in the hierarchy of the New York Central Railroad, including members of the Vanderbilt family. In due course my father married the daughter of Emily Vanderbilt and William Douglas Sloane. Emily was the granddaughter of Commodore Cornelius Vanderbilt and the daughter of William H. Vanderbilt. And so the commodore's granddaughter is my grandmother. She was a grand dame and, as far as we could remember, a rather stern lady. My mother took us dutifully to visit her once a month or so, and it was never a terribly exciting or pleasant occasion. We went through it and she was kindly. Commodore Vanderbilt, my great-great-grandfather, established the New York Central Railroad and his son, W.H., my great-grandfather, later took over control of the railroad from his father and made it even more successful.

On the other side of my mother's family, my mother's father was William Douglas Sloane (fig. 3). His father, my great-grandfather Sloane, came from Kilmarnock, Scotland, and was the originator of the W. and J. Sloane Carpet and Furniture Store in New York City.

My mother, Lila Vanderbilt Sloane, was the youngest of three daughters. There were also two sons, but one died in infancy, and the other one, Malcolm Sloane, died when he was thirty-five or forty. My mother and father met at a big dinner party hosted by her parents. They became seriously interested in each other, but there was a rule in the Vanderbilt family that no one could marry until they were twenty-three years old, so my parents were married in July 1902, the year my mother turned twenty-three (fig. 4).

My mother, Mama, was prettyish I would say (fig. 5). What I remember most was that she was always complaining about having to lose weight. Mama was a very kind, very sporting lady. She skated, snowshoed, played tennis and golf—she was a good rugged lady. She certainly was not what one would call an intellectual, but had a lot of horse-sense. She learned languages and certainly could do arithmetic because she kept the books for the family, but she was not wildly enthusiastic about museums and such. This bothered Papa. Mama was always very hospitable to his friends in the art and theater world, but knew and cared very little about the arts. Mama used to get dreadfully seasick but Papa finally got her to Europe to show her the sights. Years later Papa took my brother and me to Europe and at a museum somewhere in France, he said, "Well, here is where your mother just broke down and said 'I just can't go through another museum.'"

FIGURE 4

Mama liked people and people liked her. Her interests were reasonably simple—she was concerned with her family and her friends. She never put on the airs of a grand lady or particularly liked the glitter.[2] I don't think she was particularly concerned with politics, but she was interested in what was going on internationally, and spoke both French and German. She also was very sensitive to social problems and did what she could to help those in need.

Early Childhood

I was born in my Grandmother Sloane's house on January 30, 1904, at 2 West Fifty-Second Street. It's now of course the site of a big commercial office building. In about two years, we moved into a private residence around the corner, 645 Fifth Avenue, where I grew up. My family was not terribly rich like some people, but my parents provided quite a nice home.

My brother Fred was born fourteen months after I was (figs. 6–10). He and I were close in those days, although he always thought I was the favorite of the family and I always thought he was the favorite. I had two sisters, both younger than I. My brother and I thought they profited from our experiences because they were born four or five years after we were born. From our point of view they had the great benefit of our enlightenment. The older of the two was Marjorie, whom I helped name. I remember when Papa told me that I had a baby sister, I had just come across the name Marjorie in a book I had been reading, and I thought it was quite a nice name. So I asked,

FIGURE 4.
My parents Lila Vanderbilt Sloane and William Bradhurst Osgood Field on the occasion of their engagement. (Probably taken in 1901; photo courtesy J.O. Field.)

FIGURE 5.
Lila Vanderbilt (Sloane) Field. (No date; photo courtesy J.O. Field.)

FIGURE 5

FIGURE 6.
"William Osgood Field about 1910, 6 years old." (No number; courtesy J.O. Field.)

FIGURE 7.
My father William Bradhurst Osgood Field with my brother Fred Vanderbilt Field (R) and me (L) about 1910. (No number; photo courtesy J.O. Field.)

FIGURE 8.
Fred (R) and I in our wagon, June 1911. Photo taken by our father. (WF-M-11–1; photo courtesy J.O. Field.)

FIGURE 9.
At High Lawn in January 1911. From left to right my mother Lila V. Field, my brother Fred, my grandmother Augusta C. Field, probably our nurse Dolly Kernochan, and me. (WF-L-H-2; photo courtesy J.O. Field)

FIGURE 10.
An early photograph taken by Bill. His grandmother Augusta Curry Field ("Mia") is in the car. A note on the back, written by his grandmother reads, "Taken by Little Osgood May 11, 1912 on Riverside Drive where our motor stood when we went down to our sacred garden. He had it enlarged—his own plan and when I offered to pay for it he said, "'No Mia, not if it cost me ten dollars instead of 50 cents.'" [Ed note: Bill was called Osgood or Os by his family and friends in his early years, but he changed to Bill during World War II.] *(Photo courtesy J.O. Field.)*

FIGURE 6

FIGURE 7

FIGURE 8

FIGURE 9

FIGURE 10

FIGURE 11

FIGURE 13

FIGURE 12

FIGURE 14

FIGURE 11.
My mother with Marjorie sitting on her right and baby May in her left arm. Taken around 1912. (No number; photo courtesy J.O. Field.)

FIGURE 12.
My sisters May (L) and Marjorie (R) in 1915. (WF-HL-15-109; photo courtesy J.O. Field.)

FIGURE 13.
My sisters and brother about 1914. I am on the left, then my sister Marjorie, sister May, and Fred on the right. (No number; photo courtesy J.O. Field.)

FIGURE 14.
When my father was stationed in Ithaca, New York during World War I, we occasionally went to visit him. I took this picture in 1918. (No number; photo courtesy J.O. Field.)

"Why don't we name her Marjorie?" And she was named Marjorie Lila Field, Lila being my mother's first name. Sixteen months later my second sister Mary Augusta Field was born. Augusta was my Grandmother Field's first name, and Mary was my father's sister. We called her May (figs. 11–13).

My sisters look like my mother, but were tall like the Field family. I look more like my dad, with my blue eyes and the shape of my nose. Papa was a good-looking man, quite an attractive gent. He was a big man, tall and around 200 pounds, strong and with a good presence. My brother and I were small in comparison. I weighed only 135 pounds in high school. In college I decided to put on weight, so I went out for crew and rowed with the 150s [150-pound team] for three years. It's hard work , but it's the satisfaction of eight people pulling oars together and feeling the boat surge as the oars dip into the water.

My early life was a pleasant, if rather unproductive, life and as I look back, I spent a lot of vacations playing tennis or golf and maybe reading. Later I chose not to continue this sort of existence. I did not approve of the contrast I saw between the way people lived in those days. That, of course, became very much accentuated during the depression, which I remember vividly, and which changed my way of looking at things. And my values. I've never prized money. As long as you're comfortable and can operate and try to be useful to society, that is the main purpose. On the other hand, my family situation also made it possible for me to be independent, and not have to go into business, at which I wouldn't have been good anyway. I was much more satisfied working for a nonprofit organization than for one whose primary objective was to make money. My brother had the same feeling about society that I did, but he took a

FIGURE 15

FIGURE 16

FIGURE 17

FIGURE 18

FIGURE 15.
Engine 8 in New York City in 1916. (F-NY-16-24; photo courtesy J.O. Field.)

FIGURE 16.
An old steam fire engine with a motor front drive in New York City in 1918. (F-NY-18-7; photo courtesy J.O. Field.)

FIGURE 17.
A ladder truck in the Pittsfield, Massachusetts parade in 1915. (F-HL-15-54; photo courtesy J.O. Field.)

FIGURE 18.
A steamer on parade in Lenox, Massachusetts, 1920. (F-HL-S-51; photo courtesy J.O. Field.)

somewhat different course. He went into left-wing activities and has written an autobiography.[3] He thought he was helping the world, but now there is some question as to whether it brought any long-term relief or happiness.

During World War I my father joined the army and served in intelligence (fig. 14). I think that was the happiest, most satisfying time in his life because he had something to do, a real and an important job. He was stationed first at Cornell University, where the U.S. Army Signal Corps had a school for the interpretation of aerial photography. He was very interested in this. He then went from Cornell into military intelligence in Washington. In those days, the intelligence section consisted of a general, who was an ex-West Pointer, my father, who was a captain, and a secretary. Those three as I recall were the extent of intelligence during World War I. We were all very proud of him.

Fire Engines and Railroads

My interests as a child were such things as fire engines and railroads, and Papa cultivated those interests. In those days the fire engines were drawn by horses, and it was always quite a sight to see them come by with the dog in the lead,

barking. Papa knew the fire department chaplain in New York City and told him that I was interested in going to a fire. Shortly after that, the chaplain took me in this horse and carriage to a warehouse fire in the Lower West Side. And of course, we had to go through the fire lines. The fire was pretty well under control by the time we got there, but I remember seeing the firemen who had been overcome by smoke laid out on the sidewalk. The fire was soon out and we left, but it was quite a thrill. I still take pictures of the apparatus. I've seen the change from the old horse-drawn steam pumpers to the present diesel-operated pumpers. I guess this interest is something that will always be with me (figs. 15–18).

Papa also cultivated my interests in the railroad—he knew a lot about its operation. He told me of his various inspection trips on some of the trestles while working for the New York Central. A train would come along while he was on a bridge and there was nothing to do but to lower himself down and hang on by his hands while the train went by above, and then he'd get back up on the track. Those must have been interesting days. Well, this gave me an interest in railroading that I have always had, and I still follow all the new types of equipment, although I loved the steam era best (fig. 19).

ST. NICHOLAS
THE LETTER-BOX
XLII   SEPTEMBER, 1915   No.

Copyright, 1915, by THE CENTURY CO. All rights reserved.

TWO RAILROAD HEROES

"SAY, Jack," said Tom the fireman to the engineer of No. 4 of the westbound train on the Leadville and Hayden Railroad, "did you see the dam up at Twin-lakes is getting weak, and they are afraid it will break?"

Jack looked at his watch and announced that the train was due.

As they were leaving the room, Tom said to Jack: "You must be careful to-night, the tracks are very slippery, and it is snowing hard."

When they got on the engine, the train pulled in. After the other engine had run into the yard, Jack and Tom took their places. When the signal was given, Jack opened the throttle slowly and the big locomotive started on its way to Hayden, about 109 miles distant.

On they sped through the night till they arrived at a tunnel. After coming out of the tunnel, they heard a peculiar roar that came from the cañon below them. They were beginning to cross the trestle when the noise came louder and louder; finally, just about in the middle of the trestle, Tom, peering through the darkness and snow, cried:

"Open the throttle, Jack! there is a big lot of water coming! It will wash the trestle away."

Jack opened the throttle and the engine bounded forward. But just as the train had stopped, there was a furious blast of broken wood and the trestle had fallen.

After stopping to see if all was right around the train, they started back towards the engine. Jack said:

"Come on! The freight that leaves Leadville in fifty-two minutes will be wrecked if we do not hurry and get to Hayden! We won't be able to telegraph back."

So they started at full speed and got to Hayden ten minutes early, two minutes before the train was to leave Leadville. Then they telegraphed back and so saved the freight.

WILLIAM OSGOOD FIELD (age 10).

1054

FIGURE 19

FIGURE 20

High Lawn

The family had a fairly routine existence when I was growing up. From about January first till sometime around the end of school, we lived in New York City. Then in the summer, we'd move to the country, to High Lawn Farm in the Berkshire Mountains of Massachusetts, and stay there until the end of December. So in my childhood I went to school both in the country and in the city.

High Lawn Farm in Lenox, Massachusetts, belonged to my Grandfather Sloane. It was on a high cleared plateau and had good crops, as well as cows, chickens, and so forth but it was not a commercial operation. Beginning about the 1870s, Lenox was a place where people had gone to build summer residences. My Grandfather and Grandmother Sloane were among the first to go there. They built a big wooden house called Elm Court in 1884–85. They added to it from time to time, many guestrooms, attics that went from one end to the other and lots of little side corners where my cousins and my brother and I used to play hide and seek. It was a fantastic old place.

FIGURE 21

FIGURE 22

FIGURE 23

FIGURE 24

FIGURE 19.
"Two Railroad Heroes" was written by Bill when he was in fifth grade. The inscription on the back, probably written by Bill's mother, reads, "January 1915. A composition by Osgood during his first year at the Allen-Stevenson School. Mamaime [Bill's grandmother] sent it to the editor of the St. Nicholas Magazine quite unknown to anyone, as she felt it was too good to be lost. In due time they accepted it and enclosed to 'William Osgood Field' a check for $1.50 with a paper for him to [sign] that it would not be given elsewhere for publication." (Courtesy J.O. Field.)

FIGURE 20.
High Lawn House in the 1920s. (No number; photo courtesy J.O. Field.)

FIGURE 21.
On High Lawn Farm in 1918. (F-L-18-S-76; photo courtesy J.O. Field.)

FIGURE 22.
On High Lawn Farm in 1915. (F-HL-15-33; photo courtesy J.O. Field.)

FIGURE 23.
We played ice hockey on the pond at High Lawn during the winter. This was probably taken in 1920. (No number; photo courtesy J.O. Field.)

FIGURE 24.
I played hockey at prep school and I was on a house league at college. This picture was taken at High Lawn in 1920. (F-L-20-S70; photo courtesy J.O. Field.)

FIGURE 25.
Fred (R) and I (L) dressed for our trip into the Cave of the Winds in 1913. (F-NF-13–11; photo courtesy J.O. Field.)

FIGURE 26.
The camps had cabins, but they were nothing fancy. There were outdoor toilets and no running water that I remember. My father took this photo in 1907. (No number; photo courtesy J.O. Field.)

FIGURE 27.
My mother and father at Camp Brulé. This photo was taken in either 1908 or 1915. (No number; photo courtesy J.O. Field.)

FIGURE 28.
The guides at one of the hunting camps in 1908. The guides were mostly Scotsmen who had settled in New Richmond, an interesting little community on the Gaspé Peninsula. There was always a certain amount of antagonism between the Scots and the French, as you'd expect, but they managed to get along. Photo taken by my father. (No number; photo courtesy J.O. Field.)

FIGURE 25

FIGURE 26

FIGURE 27

FIGURE 28

Meanwhile my grandfather deeded some of the land to my mother and father where they built High Lawn House in 1908–1910 (figs. 20–24). Then my grandfather died a few years later, in 1915, and my mother and father inherited the rest of the farm. It then became one unit and my sister Marjorie and her husband still operate it that way, living at High Lawn. High Lawn Farm now is well known in the community for its Jersey cattle, so my sister and her husband have made quite a success of this operation. Elm Court is still there but it hasn't been used for years and has fallen into disrepair.

We did a certain amount of work around High Lawn when I was a boy, but very little as I remember. Mama loved to work in the garden and used to make faint efforts to get us to help her. But, early in life, I made the pretence that I couldn't tell a weed from a vegetable from a flower that was to be cultivated. To this day, I still play dumb when it comes to weeding.

Family Trips

My mother and father were very close to us when we were young. They were a lot of fun and took us on many trips, although we did not travel to any place very far away. Mama got seasick very easily, so she was not keen about going across the Atlantic. Papa used to say that, at some point before we were born, they took a trip to Europe. They got aboard the boat in the evening, and during the night she got violently ill. It turned out they hadn't left the dock yet! The first big trip that I recall was to Niagara Falls, about 1912. We went into the Cave of the Winds, under the American Falls. You walked on the boardwalk behind the falls and you got soaking wet even though you were specially dressed for that (fig. 25). As a result of getting soaked I had the croup for a couple of nights. I remember that very distinctly.

Mama was quite an outdoorswoman and I know that influenced me. She loved the outdoor life, fishing and hunting. She inherited that from her aunt, Mrs. Seward Webb [Lila Vanderbilt Webb] who had been out to Wyoming in the last century. Papa had a schooner in Florida and every winter my father and mother would go there tarpon fishing. Mama liked fishing but still hated rough seas. For a while my mother had the record for a woman's catch, 200 pounds or so. I grew up with it hanging on the wall.

My mother and father also went on some hunting and fishing trips to Québec and New Brunswick around 1907–1910. They belonged to a little club of twelve members, mostly

FIGURE 29

FIGURE 30

FIGURE 31

FIGURE 32

FIGURE 33

FIGURE 34

FIGURE 35

FIGURE 29.
My mother with her catch, 1908. (No number; photo courtesy J.O. Field.)

FIGURE 30.
My father always did some hunting while at Camp Brulé. This was 1907. (No number; photo courtesy J.O. Field.)

FIGURE 31.
I first went to the Little Cascapedia in 1915 with my mother, father, and brother. We are on the bottom steps of the cabin at Camp Brulé and the guides are on the upper step. (F-LC-15-2; photo courtesy J.O. Field.)

FIGURE 32.
We had to pole up the rapids to reach the fishing spots. There were two men in a boat, one poling in front and one poling in back, and normally the passenger, in this case my mother in 1908, would sit in the middle. (F-LC-08–11; photo courtesy J.O. Field.)

FIGURE 33.
The train station at New Richmond. My father took this probably in 1907 or 1908. (No number; photo courtesy J.O. Field.)

FIGURE 34.
My mother (center) and possibly my grandfather on the New Richmond station platform. Photo taken by my father probably in 1907 or 1908. (No number; photo courtesy J.O. Field.)

FIGURE 35.
We were met by a carriage and taken up the road to the first camp. My father probably took this photo in 1907 or 1908. (No number; photo courtesy J.O. Field.)

FIGURE 36.
Looking down Fifth Avenue in New York City. Taken by my father probably in 1915 from in front of our house at 645 Fifth Avenue. (WBOF-NY-10/15–4; photo courtesy J.O. Field.)

FIGURE 37.
The view up Fifth Avenue from near 47th Street taken by my father probably around 1910. (No number; photo courtesy J.O. Field.)

FIGURE 38.
The second Madison Square Garden, Madison Avenue at 26th Street. This was taken by my father in 1910. (WF-NY-10-L4; photo courtesy J.O. Field.)

FIGURE 36

FIGURE 37

FIGURE 38

Canadians, each member owning shares in three camps along the Little Cascapedia River, up in the Gaspé Peninsula (figs. 26–31). The Grand Cascapedia and the Restigouche were the great salmon-fishing rivers in the days before I went up. My parents took Fred and me up several times after my sisters were born. The boats we used were somewhat like the old Adirondack Guide Boats—round bottomed, long and sturdy. They could hold a heavy load and they didn't turn over easily in the rapids (fig. 32). I used to do some poling in the front. We'd go up river in the morning and come back in the evening. Those were very pleasant trips. I was not an enthusiastic fisherman—I didn't have the patience—but I loved the outdoor life and the work on the river. Soon, I became more and more interested in going way up river, just to see what was up there.

Getting to the camp was always interesting (map 1). We took the night train to Montreal, spent the day there, and boarded another night train heading to Halifax, Nova Scotia. Early in the morning, we'd get off the train at Matapedia, cross a platform, and get on a train with little engines and cars that looked as though they'd been on the New York elevated railways. This equipment was from at least thirty years earlier. This train ran about 150 miles up the coast of the Gaspé Peninsula and we'd get off in New Richmond (figs. 33–35). We loved this sort of life.

Formal Education

From first through fourth grade, I went to the Old Lenox Academy, called Trinity School then, for the fall term, September through December, while we were at High Lawn. It was a little school—I suppose there were about thirty students. During the winter and spring, January through May, I went to Charlton School in New York City. From fifth through eighth grade, I went to the Allen-Stevenson School at 50 East Fifty-Seventh Street in New York all year. During the fall, the family took the Sunday afternoon train from Lenox to New York, and came back Friday afternoon. I remember we used to rollerskate up and down Madison Avenue, going back and forth to home for lunch. There was much less traffic in those days, and of course it moved much slower (figs. 36–38). Winters were sort of a chore. We lived about eighteen blocks, which is almost a mile, from Charlton, and seven blocks from Allen-Stevenson. There were some cold walks in the morning up against the north wind but somehow we survived. The clothes that we wore in those days would be grotesque today—layers and layers, great heavy coats, and boots that nowadays would be museum pieces.

For high school it was customary for my mother's friends and relatives to go to Groton and St. Marks and similar private boarding schools in eastern Massachusetts. My father heard about Hotchkiss, a new prep school with a good reputation and in due course I joined the prep class there in September 1917. I always had difficulty with foreign languages, particularly German and Latin. In my senior year, it became a question of working all summer and being tutored to make up, or to take a fifth year. Wisely, we decided I would take a fifth year. I was young, I wasn't ready to go to college. So I took a fifth year at Hotchkiss and graduated with the class of '22. And I've never regretted that. I was a year older, more mature, and ready to take on college. That fall I entered the class of '26 at Harvard.

To my surprise, in the middle of that first fall at Harvard, the dean called in the four of us from Hotchkiss and said, "We'd like you to come to lunch. You four have won the honor of having the highest grade average of any group of four or more coming from any prep school." So we got a plaque and Hotchkiss had a holiday in our honor. That's the only academic honor I think I ever received.

In my first year I took the general course in geology taught by Reginald Daly,[4] a well-known geologist. He was an extremely good lecturer, very enthusiastic, and got you excited about rocks and erosion and other geologic processes and features. Due to the interest he aroused in me in that course, I decided to major in geology. I liked mountains and rivers and waterfalls and alpine environments. The study of geology in the branches now known as geomorphology and physical geography brings one right into those subjects, and you understand a little more about what you are looking at. In due course, with considerable strain, I graduated in June of '26.

The Canadian Rockies

*We were not pioneers ourselves,
but we journeyed over old trails that
were new to us, and with hearts open.
Who shall distinguish?*

J.M. Thorington, 1925

After World War I we took more trips than just to eastern Canada. Papa visited the Canadian Rockies (map 2) in 1919 with his friend Dr. Paddock, the local doctor of Pittsfield, and came back with photographs and talked enthusiastically

FIGURE 39

FIGURE 40

FIGURE 41

FIGURE 42

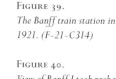

FIGURE 39.
The Banff train station in 1921. (F-21-C314)

FIGURE 40.
View of Banff I took probably in 1920 showing Cascade Mountain and Banff Avenue.

FIGURE 41.
On several occasions we stayed at the Canadian Pacific's Chateau Lake Louise. I took this photo in 1921, and in July 1924 the wooden wings of the Chateau burned to the ground. By 1925 a new concrete wing was built and the hotel remained just about the same until major remodeling began in the early 90s. (F-21-C208)

FIGURE 42.
My father in camp in 1920. (No number; photo courtesy J.O. Field.)

FIGURE 43.
I did some hunting on those early trips to the Canadian Rockies, but I was becoming more interested in seeing the countryside. (F-22-C204)

FIGURE 44.
Our camp at Bow Lake in 1924. (F-24-458)

FIGURE 45.
Crossing the Saskatchewan River in 1924. This is now the site of the bridge on the Lake Louise to Jasper Highway. Jim Boyce was our outfitter and guide all three years, 1920, '21 and '22. He was an extremely good leader and a very pleasant, interesting, and fine man. (F-24-56)

FIGURE 46.
Looking south across the Saskatchewan River in 1924 to what was called White Pyramid (see arrow), now officially named Mt. Chephren. (F-24-59)

FIGURE 43

FIGURE 44

FIGURE 45

FIGURE 46

about the trip. So, due largely to our pressure on him, Papa took my brother and me there in August of the next year, 1920, and again in 1921 and 1922. Those were four- to six-week trips. I was sixteen years old on the first trip. We were looking for animals, possibly to hunt some other year, taking pictures, and seeing some new country. The Canadian Rockies were my first views of high mountains and from the very beginning I was fascinated by their high waterfalls, deep canyons, and vast fields of snow and ice. After leaving Calgary on the Trans-Canada Express at eight or nine o'clock in the morning, we could see the sharp rocky peaks from the window of the train. We were due in Banff at noon, so the last three hours approaching and entering these mountains was a great thrill that I've never forgotten.

That first trip we got off the train at Banff and spent a night in town (figs. 39, 40). Then we went six or seven miles out of Banff towards Lake Louise (fig. 41) to where the horses for the pack train were kept. Hiring an outfitter with a pack train was the only way to go into places otherwise inaccessible. We took the horse trail up over the mountains and passes between Banff and Lake Louise (map 2): up over Cascade Creek, north to the Red Deer, which is quite a big stream, and from there to the Clearwater, up the Clearwater, and over Clearwater Pass—one of the high ones—then over Pipestone Pass and down the Pipestone back to Lake Louise, from where we took a train back to Banff. It was a good introductory trip. The mountains along our route weren't the highest, but it was pretty country. On that trip I saw my first glaciers, off in the distance.

The second year, 1921, we went from Lake Louise up the Pipestone into Clearwater country to points outside Banff National Park, to do some hunting (map 2; figs. 42, 43). I don't look back on it with any great satisfaction or pride, but I think there was a different attitude in those days about trophy hunting. It was the thing to do back then. You rationalized it by looking for the old rams, or the old moose that might die a slow death the next winter, and so you saved them from that by shooting them down that September. I've changed now. Other than maybe a snake or a skunk, I haven't shot an animal since probably 1928.

On the 1922 trip the next year, we went up to the Columbia Icefield (map 3). At that time there was no highway between the icefield and Lake Louise—that was built in the '30s.[5] It was about five or six days' travel by pack train from Lake Louise, averaging eighteen to twenty miles a day. And I still think in terms of that, even though now you roll along the highway at sixty miles an hour and easily cover twenty miles in twenty minutes. It was two days to Bow Lake (fig. 44). It was then probably two days down to the Saskatchewan crossing, which you planned to cross early in the morning when the water's lower (figs. 45–48).[6] One more day to go up the North Fork of the Saskatchewan, and then the next day would take you to the vicinity of Athabaska Glacier (map 3). The glacier came right down into the woods, so the regular route on to Jasper was over Sunwapta Pass, the next valley behind the present-day Columbia Icefield Chalet, which of course wasn't there then. We camped at what was known as Camp Parker. One day, our outfitter Jim Boyce took us up to Parker Ridge and we looked down on the Saskatchewan Glacier. I took two pictures, one looking down on the terminus of the glacier (fig. 49) and the other looking up the glacier toward Mount Castleguard. Those were the first pictures of an identified valley glacier I ever took. I had no thoughts other than that it was interesting and spectacular. The next day we went up to Wilcox Ridge that stands behind the present Columbia Icefield Chalet and looked down on the Athabaska Glacier. I took some pictures in bad weather (fig. 50), so the picture of that glacier in 1922 is not a good one, but good enough to identify the landmarks around the terminus.

Setting the Stage

Those trips to the Canadian Rockies had an effect on my life which wasn't obvious at the time. This was really the beginning of a conscious effort to photograph glaciers from a known position. The combination of my interest in photography and my interest in looking for changes in glaciers later developed into my lifelong work. It began in the Canadian Rockies, and I look back on those trips with great fondness. We came back home full of exuberance, and a love of travel, of the mountains, and the fun of taking pictures. I can remember hours of boring the family friends with photos of the Canadian Rockies, year after year.

The Alps

In the summer of '23, after the three trips to the Canadian Rockies, Papa took my brother and me abroad. I had just finished by freshman year at Harvard and a friend of mine in the same dormitory, Lement Harris, had a picture of the Matterhorn in his room. His father had taken him to Switzerland the previous summer, and told him that if he returned the next summer

FIGURE 47

FIGURE 48

FIGURE 49

FIGURE 50

FIGURE 47.
It was never an easy job to get the horses across the river. They would wander off into deeper water and you see how the bottom of the packs would get wet. Somehow it was always the toilet paper that was packed in the bottom. My wife says that I always carry toilet paper in my pocket to this day. (F-22-C101)

FIGURE 48.
These were some of the Indians we would meet along the way. These probably are the Beavers in 1922, a Stoney Indian family at Kootenay Plains. (F-22-C116)

FIGURE 49.
Saskatchewan Glacier from Parker Ridge, 1922. (F-22-153)

FIGURE 50.
Athabaska Glacier from the south ridge of Wilcox Mountain, 1922. (F-22-158c)

FIGURE 51.
We stayed at the Hotel Monte Rose in Zermatt. (F-23-845)

FIGURE 52.
The town of Zermatt in 1923 with the Matterhorn in the distance. (No number.)

FIGURE 53.
Our guide Alex Graven (left), Fred and myself (right) at the Zermatt railroad station. Even today the only way to reach Zermatt is by train. (F-23-939)

FIGURE 54.
I enjoyed taking pictures of the various types of traffic in front of the hotel. (F-23-863)

FIGURE 55.
Some of the streets and houses of Zermatt were very interesting. [Many such buildings were still there in 1985.] (F-23–934)

FIGURE 56.
Emil Graven (left), our guide on several climbs, and his sons, Alex and Aloise (right), our guides on all the climbs. (F-23-824, 825)

FIGURE 51

FIGURE 54

FIGURE 52

FIGURE 55

FIGURE 53

FIGURE 56

FIGURE 57

FIGURE 58

FIGURE 59

FIGURE 57.
The Matterhorn from the summit of the Riffelhorn. (From a picture purchased in Zermatt, 1923.)

FIGURE 58.
On the summit of the Matterhorn. Myself (L), Fred and Emil Graven (R). (F-23-709)

FIGURE 59.
Climbing the Rifflehorn. Aloise Graven in the lead followed by myself, Fred, then Alex Graven. (F-23-554)

We stayed on in Switzerland for another ten days or so and climbed several other peaks (fig. 59). So we got a taste of mountaineering that summer. This was also the first time I ever set foot on a glacier. I still see Lement every year at our class of '26 reunion dinner in New York, though he's taken a different path in life than I. He came from a banking family and got fed up with that sort of lifestyle. He saw other things that he thought were more important in life, and became an avowed Communist.[7] He still is as far as I know. We don't go deeply into these things. He's the only Communist that I ever knew who had a sense of humor. He can make wisecracks about it, but all the others I knew were terribly serious.

First Ascents

In '24 my brother, my friend Lement, and I planned a trip to the Canadian Rockies to do some climbing. We were intrigued by the first ascents still available. You could read about the early first ascents—they were very dramatic.[8] Most of the highest peaks had been climbed, but there were still a couple of pretty good peaks over 11,000 feet unclimbed. We were not expert mountaineers, but we could follow a guide most anyplace. I had made arrangements with the Canadian Pacific to hire Ed Feuz, Jr.,[9] as our guide. I think we paid him five dollars a day. Lement's father wanted him to have the Swiss guide with whom Lem had climbed in the Alps the previous summer—his father had said that he couldn't go anywhere without Joseph Biner. So he brought Biner over. That was the way things were done in those days (figs. 60, 61).

he would take him up the Matterhorn. Well, my reaction was, if you can climb the Matterhorn, I can too! So, as a reward for going through endless museums, cathedrals, and monasteries in England and France, we went to Geneva, Switzerland, and then on to Zermatt (figs. 51–56). When Harris arrived we climbed the Matterhorn (figs. 57, 58). But I decided that, instead of descending the standard route to Zermatt via the northeast [Hornli] ridge, I would make the traverse with my guide and go down the Italian side. According to what I had read, it was longer but not too difficult. I parted company with Lement, my brother Fred and their guides on the summit and I went on down the Italian side with my guide. We joined a Viennese doctor on the way down, who told us his friends went to the Alps each year and climbed something a little more difficult than the previous year with the idea of eventually climbing the Matterhorn. Well, he decided to do the Matterhorn first instead of wasting all that time. So we went down, both equally inexperienced.

FIGURE 60.
The Swiss guides Joseph Biner from Zermatt (L) and Ed Feuz from Lake Louise (R). (F-24-128)

FIGURE 61.
A postcard given to me by Ed Feuz.

FIGURE 62.
Watchman Peak (at arrow) from our camp at Castleguard Meadows. It isn't very high, 9872 feet, but it is a striking landmark. (F-24-101)

FIGURE 63.
We had to cross the Castleguard River up near the head on this bridge. The meltwater from all the glaciers in Castleguard Meadows above fed into the river, so it was quite a torrent, especially at the usual high water stage in the afternoons. (F-24-246)

FIGURE 64.
A cave we entered one morning is known[10] to extend some four to five miles back under the Columbia Icefield. We came back later in the day and found a torrent of water roaring out. The walls of the cave are very porous and this was the meltwater from the icefield. (F-24-206)

FIGURE 60

FIGURE 61

FIGURE 62

FIGURE 63

We hired Max Brooks, Jim Boyce's partner, again as our outfitter and chef, and Ernie Stanton and Cecil Smith as our packers to take us up to the Columbia Icefield (map 3). On that trip we camped at Castleguard Meadows (figs. 62–65). I think we made about five or six first ascents (map 4). One was a good high one, South Twin, one of the main peaks in the Columbia Icefield area (figs. 66–70). The climbs weren't so technically difficult, but they were long. You left camp in the evening, and you walked all night while the surface of the glacier was frozen and made good walking. You'd reach the summit about sunrise, and then start back. I think one round trip took us twenty-eight hours and another about twenty. But we had a good time. On our way back to Lake Louise, we stopped at Bow Lake where Jim Simpson was just finishing the first structure of his lodge Num-ti-jah (fig. 71). I made one last climbing trip to the Canadian Rockies in 1925, which I described in the Harvard Mountaineering Club publication.[11]

FIGURE 64

FIGURE 65

FIGURE 66

FIGURE 67

FIGURE 68

FIGURE 69

FIGURE 65.
Henry Schwab (L) and Henry Hall (R), both prominent members of the American Alpine Club (AAC), were passing through the area and stopped by our camp. Henry later put me up for membership in the AAC. Sir James Outram was also in the party, his first visit to the area since his great expedition of 1902 when he made first ascents of many of the high peaks in the region. (F-24-238)

FIGURE 66.
South Twin (L) was the highest unclimbed peak in the Canadian Rockies at that time, 11,675 feet. North Twin (R) is the third highest in the Canadian Rockies, 12,085 feet. (H. Palmer photo, taken in 1924. Copy photo in the WOF Collection. Palmer Collection at the Whyte Museum of the Canadian Rockies, Banff, Alberta.)

FIGURE 67.
The party on the summit of South Twin. Left to right: Lem, Feuz, Fred, Biner. We also climbed North Twin, but it was a second or third ascent, not a first. (F-24-143)

FIGURE 68.
Back in camp after both ascents, the party with our "Twins" banner. Left to right: Max Brooks, Joseph Biner, Lem Harris, Cecil Smith, Fred Field, Ed Feuz, and Ernie Stanton. (F-24-161)

FIGURE 69.
We also climbed Mt. Castleguard, 10,096 ft. The party included Biner (L), Fred, Brooks (seated), Lem and Feuz (R). (F-24-192)

FIGURE 70.
Mt. Columbia, 12,294 feet, is the second highest peak in the Canadian Rockies after Mt. Robson and, I believe, drains directly into three oceans. We made the fourth ascent of this peak. From the summit, we could see only faintly the bottom of the valley below because of the smoke from forest fires in British Columbia. This view is from South Twin. (F-24-146)

FIGURE 71.
The first structure of Jim Simpson's at Bow Lake, July 1924. (F-24-464)

FIGURE 72.
Standing on the platform of Glacier House, August 18, 1893 prior to setting out for Mount Fox are the porters Demster and Stables (in back) with C.S. Thompson, H.P. Nichols and S.E.S. Allen. Upon their return, Reverend Nichols preached on "The Glory of Aspiration." (J.H. Scattergood photo, courtesy AAC Archives.)

FIGURE 70

FIGURE 71

FIGURE 72

The American Alpine Club

As a result of the 1923 season in the Alps, and the 1924 trip to the Canadian Rockies, I rated membership in the American Alpine Club (AAC). The club had been formed about 1902 and it was still fairly small, with only several hundred members by the 1920s. It is likely that Henry S. Hall, Jr., whom I met while I was at Harvard, put me up for membership. We used to go to his house in Cambridge once in a while. He was a very distinguished member of the AAC. He was very enthusiastic about climbing and especially wanted to interest young people in it. He advised us on mountaineering, including the trips to the Canadian Rockies and the '23 trip to the Alps. Harry P. Nichols, an Episcopal clergyman, who was then president of the AAC, seconded me for membership (fig. 72). I became a member in 1924 and have stayed a member all these years.[12] I was now becoming fully aware of my love of travel and exploration, of the mountains and of photography, and took advantage of every opportunity that came along to pursue these hobbies.

Notes

1. See Preface for an explanation of the photo numbering system used by Bill (and to a certain extent by his father), especially how to determine the date from the photo number.

2. Bill's brother Fred described their mother very well in the following excerpt from his autobiography *From Right to Left: An Autobiography* (1983, 8–9):

    My mother devoted herself to her own mother, her husband and her children, all in a traditional, conventional way . . . . She was simple in an atmosphere of very considerable luxury. Living mostly among high-society show-offs . . . she had not a trace of vanity . . . she was a loyal friend, both to her social equals and to the people who served her. She was tolerant rather than haughty, democratic rather than autocratic.

3. Field, F. V., *From Right to Left: An Autobiography* (1983).

4. Reginald Aldworth Daly (1871–1957) was Sturgis-Hooper professor of geology at Harvard from 1912 to 1942, after which he retired as Sturgis-Hooper professor emeritus. From 1901 to 1906 he was engaged in surveying the international boundary between the U.S. and Canada for the Canadian Boundary Commission (see Appendix E for a brief description of the boundary survey work).

5. Construction of the Banff-Jasper Highway, later called the Icefields Parkway, was begun in 1931 and declared open on July 1, 1940.

6. The Saskatchewan River is fed from melting snow in the mountains. During the day, as the air temperature rises, more snow melts, raising the level of the river. Thus, one wanted to cross the river as early as possible while the water level was low and the danger the least.

7. Harris, *My Tale of Two Worlds* (1986).

8. See: Stutfield and Collie, *Climbs and Exploration in the Canadian Rockies* (1903); Monroe, *The Glittering Mountains of Canada* (1925).

9. Ed Feuz, Jr. (1884–1981) is credited with more first ascents in the Rockies than any other climber. He first came to Canada from Switzerland in 1903. He worked as a mountain guide for the Canadian Pacific Railroad from 1905 until 1949. See "Memories of a Mountain Guide," a story by Feuz as told to Imbert Orchard in Patton, *Tales from the Canadian Rockies*.

10. See the special volume of *Arctic and Alpine Research* (15, 4: 1983).

11. Field, W.O., "In Search of Mount Clearwater" (1927).

12. Thorington died in November 1989, so Bill Field was the longest living member until his death in June 1994 (Dee Molenaar, personal communication).

FIGURE 73

CHAPTER 2

SETTING FOR A
LIFE'S WORK
1925

Alaska!

In 1925, I planned my first trip to Alaska. Some cousins had been up there and spoke highly of the interesting area and of the beautiful scenery. Two friends of mine went with me. One was Harold J. Coolidge of Boston, who had graduated ahead of me in college. He was interested in wildlife and conservation, and later became quite involved in the conservation movement. The other man was Charles Dana McCoy from Texas, a friend whom I met while at Hotchkiss. McCoy was a senior at Yale that year. So the three of us took the train across the continent to Seattle. Then, on a Friday or Saturday morning, the weekly Alaska steamship left from Pier 2 down on the Seattle waterfront and headed up the coast along the Inside Passage (map 5). Three steamers rotated on the job, the S.S. *Alaska*, the S.S. *Yukon*, and the S.S. *Northwestern*. We were on the S.S. *Alaska* for that trip (figs. 73, 74). Ketchikan was the first stop (figs. 75, 76). Wrangell was the second stop

and Juneau the third. Each stop was usually two or three hours long. Sometimes if it was the right time of day, we went to the movies. Then the ship went out through Icy Strait, and headed north along the coast of Alaska to Cordova.

During the time it took to unload and load the ship in Cordova (figs. 77–79) a special train made a run for the ship's passengers up to Mile 49 from where we could see Childs Glacier, Miles Glacier, and what was known as the Million Dollar Bridge (figs. 80–85).[1] We then returned to Cordova and the boat continued on to Valdez, where we could see the Valdez Glacier (fig. 86). It was quite prominent in the distance, up-valley from the town. The next stop was the Columbia Glacier, where the steamer went to within a half a mile from the terminus and waited around for an hour or so to see if there would be some spectacular ice falls. The tourists wanted to see that in those days, just as they do today. Then we stopped at a number of canneries. That was one of the interesting parts of those trips, stopping at

FIGURE 73.
*On the deck of the S.S.
Alaska. (F-25-864)*

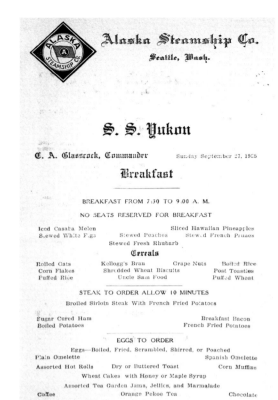

FIGURE 74

FIGURE 77

FIGURE 75

FIGURE 76

canneries in out-of-the-way inlets, to discharge or pick up cargo. The next to last stop was at the Latouche Mine (fig. 87), then on to Seward, which was the end of the run in Alaska for that steamer (figs. 88, 89). At Seward we stayed at the Van Gilder Hotel (fig. 90). It originally was built as an office building by E.L. Van Gilder in 1916, when Seward was the construction base of the Alaska Railroad, but then was turned into a hotel around 1922. It was sold about a year before we arrived and the new proprietor was an amusing fellow. From Seward we took the Alaska Railroad across the Kenai Mountains to Anchorage, which was a small town at that time, only about 2,000 people (figs. 91–95). It was primarily a division point of the railroad, where we changed engines.

On that trip we saw Trail Glacier, which terminated two or three miles from the railroad, and then past Bartlett and Spencer glaciers, both of which terminated only a few hundred yards from the tracks. These were so close to the railroad right-of-way that you wondered, if any advance in the glacier should occur, what would happen to the railroad. Indeed, the railroad was built on the 1885 moraine of the Spencer Glacier.

From Anchorage we went on to Broad Pass and Fairbanks, a trip which in those days took two days. The first day was to Curry, where there was a lodge. The next morning, if the weather was clear, you could get up at 3 A.M., climb a ridge, and get a view of Mount McKinley. It was bad weather, so we never made that trip. Then we reboarded the train that morning and arrived at Fairbanks in the evening (figs. 96–99).

One excursion took us to the Agricultural College and School of Mines, what became the University [of Alaska] in 1935. Charles Bunnell, founder and first president of the college, showed us around.[2] I took some pictures (fig. 100), but unfortunately none of him. We went to some mining sites, probably a dredge site near Fairbanks, then we took the train back to Seward.

FIGURE 78

FIGURE 79

FIGURE 80

FIGURE 81

FIGURE 82

FIGURE 83

FIGURE 78.
On the dock at Cordova. It took several hours for a ship to unload supplies for the canneries and the Kennecott Mine and to load up material from the mine. Copper ore concentrate was loaded on board the S.S. Alaska during our stop. (F-27-F591)

FIGURE 79.
Cordova, port of Copper River Railway and point from which Controller Bay coal may eventually be shipped. Town a year old. (E.A. Hegg photo, no number, July 1, 1909; copy photo from the WOF Collection.)

FIGURE 80.
The Million Dollar Bridge. (F-27-A59)

FIGURE 81.
Building the Million Dollar Bridge across the Copper River over channels NE of Long Island. McPherson Glacier in the distance. (E.A. Hegg photo #164, Aug 21, 1908; copy photo from the WOF Collection.)

FIGURE 82.
Steamer on the Copper River and a floating camp of railway laborers. Eating and sleeping on raft necessitated by the wild gorge through which the railroad crossed mountains.(Photo probably taken by E.A. Hegg, #292, 1908 or 1909; copy photo from the WOF Collection.)

FIGURE 83.
Temporary suspension bridge at the site of the Million Dollar Bridge across the Copper River at Childs Glacier. Carload of night shift workmen going to caisson to work under river level on pier of the bridge. Pipes for water, electricity, and compressed air beneath suspension bridge. (E.A. Hegg photo #272, June 2, 1909; copy photo from the WOF Collection.)

FIGURE 84.
*Miles Glacier Bridge.
September 28, 1909.
(Photographer unknown;
copy photo from the WOF
Collection.)*

FIGURE 85.
*First train over Miles
Glacier (Million Dollar)
Bridge, June 1, 1910.
(Photographer unknown;
copy photo from the WOF
Collection.)*

FIGURE 86.
*Valdez, Alaska with Valdez
Glacier in the background,
1925. (F-25-428)*

FIGURE 87.
*The Latouche Mine, a
copper mine operating on
Latouche Island in Prince
William Sound. It's now
just a relic, but you can see
where the buildings were,
and a few are still standing.
(F-25-F498)*

FIGURE 88.
*Seward was the first place
we stayed in Alaska. I'll
always have a soft spot in
my heart for the little town
of Seward, about the same
size now as it was then.
This is the main street.
(WF-27-600)*

FIGURE 89.
*The main street in Seward.
(WF-27-601)*

FIGURE 84

FIGURE 85

FIGURE 86

FIGURE 87

FIGURE 88

FIGURE 89

FIGURE 90

FIGURE 91

FIGURE 90.
The Van Gilder Hotel.
(F-25-570)

FIGURE 91.
The Hotel Anchorage.
(WF-27-608)

FIGURE 92.
On the dock in Anchorage.
(F-27-F491)

FIGURE 93.
Downtown Anchorage.
(WF-27-F606)

FIGURE 92

FIGURE 93

FIGURE 94

FIGURE 95

FIGURE 96

FIGURE 97

FIGURE 94.
Downtown Anchorage.
(WF-27-F607)

FIGURE 95.
A postcard I bought while
in Anchorage showing
a train arriving at the
Anchorage station.

FIGURE 96.
The train station in
Fairbanks in the 1920s.
(F-27-F66)

FIGURE 97.
That engine's probably still
there in Fairbanks—Alaska
Rail Road No. 1.
(F-27-F67)

FIGURE 98.
Downtown Fairbanks in
1925. (F-25-484)

FIGURE 98

FIGURE 99

FIGURE 100

FIGURE 101

FIGURE 102

FIGURE 99.
We stayed in a hotel called the Fairview, which, if it is not still there, is probably remembered.
(F-25-483)

FIGURE 100.
The old main building at the university in Fairbanks being remodeled in 1925. It was torn down a few years ago.
(F-25-497)

FIGURE 101.
Coolidge (L) and McCoy (R) on the way to Trail Glacier.
(F-25-511)

FIGURE 102.
We camped in a mosquito-infested alder patch.
(F-25-518)

We had about three days to spare before a planned hunting trip in the Kenai. We had seen Trail Glacier in the distance from the train between Seward and Anchorage and, having the necessary camping equipment, we took the train back towards Trail Glacier and were let off at Grand View Station. We walked down the embankment to the river, and about two or three miles up the valley to the glacier terminus where we camped (figs. 101, 102). One day we climbed a little peak at the head of Trail Glacier (fig. 103). Just by luck I took a picture looking down on the terminus of the glacier (fig. 104). There was a stream that came down the opposite side of the valley, just where the outermost point of the terminus rested. This is shown on the latest topo maps, so from my photo I know within a few feet where the glacier terminus was in August of 1925.[3] Then, after a three-week hunting trip to the Kenai, I returned home just in time for the opening of college. This was 1925, and that trip awoke something in me. I went into my senior year more interested in glaciers than ever.

Discovering My Predecessors

Prior to my 1925 trip to Alaska, I had heard a couple of lectures at Harvard by Reginald Daly on glaciers and glaciation.[4] And then there was Kirtley Mather, another good lecturer.[5] He'd studied the southern Andes and he was full of information about them. I also had Professor Woodworth, Jay Backus Woodworth.[6] The Woodworth Glacier in the Tasnuna area [of Alaska] is named for him. I remember his telling us that, though he didn't know exactly why the glacier was named after him. He felt he hadn't done anything that notable. He took us on some field trips to show us outwash plains, moraines, drumlins, potholes and such geological features like that in the Boston area. It is very hard to visualize these features when they are covered with houses and parkland and highways. It's after you see these things in the wild that you realize fully what they mean and the processes involved in their formation. In the mountains you are able to see the geology in front of you; it isn't hidden by vegetation or by man's developments. I'd seen

FIGURE 103

FIGURE 104

FIGURE 103.
*The cairn we built. I am on
the right and Coolidge on
the left. (F-25-540)*

FIGURE 104.
*Looking down on the
terminus of Trail Glacier.
(F-25-556)*

that in the Canadian Rockies in 1920, '21 and '22, and the trip to the Alps in 1923 enriched the experience. So the idea of changes on the earth's surface was introduced in these classes, and to me any such changes that could be recorded were of interest. If glaciers were static and unchanging, I don't think I would have been interested in them. Although the changes in the glaciers in the Alps are comparatively small, those in some of Alaska's glaciers can be hundreds of feet over a few years. And the glaciers are so accessible along the coast. In addition to Daly, Mather, and Woodworth, Charles Palache taught my mineralogy course.[7] Palache had been on the Harriman Expedition[8] to Alaska in 1899. And another professor at Harvard, Fernow, a forester, was on that expedition.[9] I knew him at some point.

In the course of hearing about glaciers from these men, plus my interest in Alaska and having seen some glaciers during my 1925 trip, I went to the science library at Harvard and asked about books on glaciers in Alaska. That is probably when I first saw Reid's report "Glacier Bay and its Glaciers," the report of his 1890 and '92 expeditions to Glacier Bay.[10] I also ran into the report on the Harriman Expedition. Volume 3, *Glaciers and Glaciation*[11] was by the geologist G.K. Gilbert.[12] In due course, I caught up with Tarr and Martin's *Alaskan Glacier Studies*.[13] Those books are classics now. They opened my eyes to what had been done back in the 1890s and early 1900s. I found that quite a lot of what these people had done had not been repeated. And the men themselves hadn't been able to go back, although

they would have liked to return. I saw that somebody had done some work, and I saw that there was a possibility of continuing it. I wanted to go back and find out exactly what happened. But at that time my plan was by no means to undertake a life's work. It was simply to try to carry out studies on the changes of the glaciers at intervals of every five years or so. When I met these people later, they were so enthusiastic about somebody continuing their observations, and their encouragement greatly influenced me to keep going back for the next sixty years.

Notes

1. From 1906 to 1910 the Copper River and Northwestern Railroad was built to connect the Kennicott Copper Mines in the Wrangell Mountains with the port of Cordova, a distance of nearly 300 km. A critical point in building the railroad was the crossing of the Copper River near Mile 49 by what became known as the Miles Glacier Bridge, also referred to as the Million Dollar Bridge because of its cost of around $1,500,000. This was a 4-span, steel structure, 470 m long at the outlet of Miles Lake, directly between the actively discharging termini of Childs and Miles glaciers. During the construction of the bridge, rapid advances of both glaciers narrowed the [distance between them] to around 3.25 km. Childs Glacier surged to within 450 m of the bridge in June 1911, and Miles Glacier reached a point

about 2,800 m away in 1910. (Field, W.O., *Mountain Glaciers of the Northern Hemisphere*, 1975: 328; also Janson, *The Copper Spike* (1979).

2. Dr. Charles Bunnell (1878–1956) was the first president of the Alaska Agricultural College and School of Mines, which became the University of Alaska in 1935. He served as president from August 11, 1921, until his retirement on July 1, 1949. See Cashen, *Farthest North College President* (1971).

3. From 1911 to 1931, Trail Glacier retreated more than 4,000 feet (Wentworth, and Ray, "Studies of Certain Alaskan Glaciers in 1931," 1936). The AGS visited the glacier in 1957 and determined it had receded an additional 1,500 feet, and it has continued to retreat since then (Field, W.O., *Mountain Glaciers of the Northern Hemisphere*, 1975).

4. See Chapter 1, note 4.

5. Kirtley Fletcher Mather (1888–1978) was a geologist and educator. He specialized in petroleum geology, glacial and regional geology, and geomorphology. He became a member of the USGS in 1910 and maintained connection with the survey until 1945. He worked for the survey in Alaska in 1923. He was appointed professor in the Department of Geology and Geography at Harvard and served as chairman of that department from 1925 to 1931 (*Who's Who in North America*, 1974).

6. Jay Backus Woodworth (1865–1925) was a geologist and one of the American pioneers in the scientific study of earthquakes. He began teaching in the Department of Geology and Geography at Harvard in 1893, and served as chairman of that department from 1904 to 1908, when he was appointed director of the Harvard Seismographical Station, which he established, and held that position until his death. He also worked as a geologist for the U.S. Geological Survey from 1918 until his death (*National Cyclopaedia of American Biography*, vol. 20, 1936).

7. Charles Palache (1869–1954) became an instructor of mineralogy at Harvard University in 1896, served as professor from 1912 to 1941, and then professor emeritus from 1941 until his death (*Who's Who in North America* 3).

8. In 1899 Union Pacific magnate Edward H. Harriman organized a crew of America's scientific and artistic elite to accompany him on his reconnaissance of Alaska, a region central to his dream of an around-the-world railway. The party, known as "The Harriman Expedition," left Seattle on the steamship *George W. Elder* on May 31, 1899, reached Bering Strait before turning back, and traveled 9,000 miles before arriving in Seattle again on July 30. The results of the expedition are published in the thirteen-volume work *Harriman Alaska Expedition* edited by C. Hart Merriam (1901–1914). Volume III, *Glaciers and Glaciation* by G.K. Gilbert (see note 12 below), is considered one of the classic books of early observations of glaciers in Alaska.

When the expedition was over, Mr. Harriman subsidized the publication by Doubleday, Page and Co. of thirteen volumes reporting on the results of the voyage (see bibliography). Volume I came out in 1901 and volume II in 1902, followed by volumes III–V and VIII–XII in 1904. Volumes VI and VII for some reason never appeared. In 1905 volume XIII was completed. After Mr. Harriman died in 1909, the Smithsonian took over publication and in 1910 reissued all the volumes with new title pages. The last volume, volume XIV, was published in 1914 in two parts. For a republication of volumes I and II together, see Burroughs et al., *Alaska: The Harriman Expedition*, 1899 (1986).

9. Bernhard Edward Fernow (1851–1923) a forester, served as secretary of the American Forestry Association from 1883 to 89, chief of the Division of Forestry of the U.S. Department of Agriculture from 1886 to 98, and director of the N.Y. State College of Forestry (connected with Cornell University) from 1898 to 1903.

10. Reid, "Glacier Bay and its Glaciers" (1896).

11. Gilbert, *Glaciers and Glaciation* (1904).

12. Grove Karl Gilbert (1843–1918) was a geologist who began his career in 1869 with the Geological Survey of Ohio and then the U.S. Geological Survey in 1879. He was considered an authority on many of the problems of the time in geology. He visited Glacier Bay in 1899 as a member of the Harriman Alaska Expedition, and was the author of Volume III, *Glaciers and Glaciation*, of the publication resulting from that trip (see Mendenhall, "Memorial of Grove Karl Gilbert," 1920: 26–44).

13. Tarr and Martin, *Alaskan Glacier Studies* (1914).

CHAPTER 3
HISTORY IN
THE MAKING
1926

My First Scientific Trip

The trip I planned to Alaska in 1926 was to be a more serious scientific effort than the trip the previous year and it would be to southeast Alaska only. We had some hunting in mind, but the primary purpose was to see what the glaciers in Glacier Bay had done since last reported by Reid in 1890 and '92 (map 6),[1] H.P. Cushing in 1891[2] (map 7), and Gilbert in 1899.[3] What is usually considered the first scientific study in Glacier Bay was by the English geologist George W. Lamplugh, who visited Muir Glacier in 1884.[4] George Frederick Wright[5] in 1886 was the first after Vancouver in 1794[6] to map the position of an ice front, the Muir, in Glacier Bay. I had read most of their reports and I had some idea of the past history of the glaciers there but not much. The area had just been made a monument the year before, 1925. Our map for the trip was the 1916 edition of the U.S. Coast and Geodetic Survey

Chart 8306 (map 9). The topography was based largely on the boundary survey work of 1907,[7] so the glacier fronts were shown as they were back then, and in some cases tremendous changes had occurred in the glaciers between 1907 and 1926. Alfred H. Brooks[8] and Arthur F. Buddington[9] had been up in 1924 and had taken a lot of photos but never wrote up anything about the trip. I knew Buddington, and probably met Brooks at some point. William S. Cooper[10] had been to Glacier Bay in 1916 and 1921 but I didn't know that prior to my trip. He had published a map of Muir Glacier and taken some photos.[11] And Mertie had also taken pictures in 1919.[12] I saw all those pictures after we came back.

A second but rather minor objective was to determine if there was a route to climb Mount Fairweather from the east. It was one of the major unclimbed peaks in Alaska. From the maps, which were not contoured, you had no concept of the gradient on the glaciers. The Ferris, which is a

FIGURE 105.
The 1926 party in Glacier Bay. Left to right: Roscoe Bonsal of South Carolina; Ben Wood, student at Harvard Medical School; Percy Pond (standing center) of the photographic studio of Winter and Pond in Juneau; Andrew M. Taylor of McCarthy, Alaska; and Paul Kegal (standing) of Juneau, skipper of the M/V Eurus, Aug 13, 1926. (F-26-451)

FIGURE 106

FIGURE 107

FIGURE 108

FIGURE 109

FIGURE 106.
Ben Wood at camp on Geikie Glacier, Aug 17, 1926. (F-26-115)

FIGURE 107.
Andy Taylor near Columbia Glacier Sta. 22, Sep 1931. (F-31-189)

FIGURE 108.
Pond's cabin Takuakhl up Taku Inlet. (F-26-64)

FIGURE 109.
The S.S. Spokane in front of Taku Glacier around 1905. (H.F. Reid photo, no number; copy photo from the WOF Collection.)

tributary of the Grand Pacific, and the Margerie looked like highways right to the upper part of the Mount Fairweather massif. Well, it didn't take more than one glimpse in person to see that those glaciers were not highways! They were heavily crevassed, with tremendous icefalls.

I should say something about the party (fig. 105). Ben Wood, who was a year ahead of me in college, in first-year medical school, joined us (fig. 106). Ben had taken an engineering course and he knew something about surveying, in which I had no experience. And another friend, Roscoe ("Rocky") Bonsal. He was from South Carolina and wanted to go to Alaska to see the countryside and maybe to hunt. He was not the least bit interested in glaciers and was cold the whole time, but he was glad to go.

For my '26 trip, I wanted somebody experienced in glacier travel. Henry Hall and others from the 1925 Canadian-American Mount Logan Expedition[13] said that Andy Taylor was the man for us. Andy (fig. 107) was from McCarthy [Alaska], and was probably the most interesting person in the group. He was very experienced in glacier travel. In the winter he carried the mail with his dog team across the glaciers and over the passes in the Wrangell Mountains, down into the Nabesna country where there was some mining operation. During the Mount Logan expedition, Andy was one of the four or five who reached the summit, and I was told that perhaps no one would have made it without Andy's help. Subsequently, Andy was with Dr. Bill Ladd and Allen Carpé and Terry Moore on the first ascent of Mount Fairweather in 1931.[14] When the team got caught in a storm and food ran short, Andy and Ladd returned down the glacier, leaving the supplies for the other two so they could make the summit, which they did. In his later years, Andy went to live with Dr. Ladd and his family and taught the Ladd children the ways of the woods.

Andy spent approximately a month with us in Glacier Bay. He was a very good companion, fun to be with. He used to say that the two books that he took with him on long trips, because they

FIGURE 110

FIGURE 111

FIGURE 112

FIGURE 113

FIGURE 114

FIGURE 110.
Taku Glacier from "Pond's Site" taken by C.W. Wright in 1904. (Wright photo #29; copy photo from the WOF Collection.)

FIGURE 111.
Rock knob in center to left of trees is "Pond's Site" (Sta. C) at Taku Glacier. Taken by C.K. Wentworth in 1931. (Wentworth photo 31-10107; copy photo from the WOF Collection.)

FIGURE 112.
Taku Glacier taken from "Pond's Site" (Sta. C) by C.K. Wentworth, June 26, 1931. The site was covered by the advancing glacier shortly after this. (Wentworth photo 31-10109; copy photo from the WOF Collection.)

FIGURE 113.
Pond had a canoe in which we paddled to landing places to view Norris, Taku, and Twin Glaciers. Ben is in the bow, I am in the middle, and Rocky is in the stern. Pond took this photo August 6, 1926 and later made it into a postcard to sell in the Winter & Pond Photo and Souvenir Store in Juneau. (Postcard from the WOF Collection.)

FIGURE 114.
Remains of the former ice-dammed lake on south side of Norris Glacier, called "Glory Lake." Note trim line extending along entire ridge behind the lake. (F-26-58)

FIGURE 115.
We went up Taku River to Twin Glacier Lake with Ben Bullard, an old man who had a farm on the east side of the lower Taku Valley about abreast of the river. I could find no remnant of the farm in later visits. (F-26-75)

FIGURE 116.
The party at Sta. A, Twin Glacier Lake, Aug 7, 1926. From left to right: Pond, Bonsal, Bullard, and Wood. (F-26-73)

Figure 117. *Paul Kegal, owner and skipper of the M/V Eurus, Aug 16, 1926.* (F-26-113)

FIGURE 115

FIGURE 116

FIGURE 117

were books he could read over and over again and they never got stale, were the Bible and the *Iliad*. I wouldn't call him an "intellectual" because I am not sure what intellectuals are, but he was a thinker, well informed, lots of fun, knew good jokes, and was a very enjoyable companion all around. On my 1931 trip to Alaska, Andy and I were alone for the first two or three weeks and I got to know him pretty well. Sometimes we'd be stuck in the tent because of bad weather, playing cribbage. He was a good cribbage player. He entertained me a lot that trip. He also was the kind of person you could depend on in a tough situation.

In Juneau we picked up the last member of our party—Percy Pond of the Winter & Pond Photo Shop in Juneau.[15] He had never been to Glacier Bay and asked to join us. We had a few extra days before we were scheduled to go to the bay, so Pond invited us to stay at his cabin up Taku Inlet (map 10; fig. 108) and have a look at the glaciers there. We went from Juneau to his cabin by the boat that serviced the Tulsequah Mine, a gold mine in British Columbia and, as I remember, the load was primarily dynamite. From Pond's cabin there was a very good view across the inlet of both the Norris and Taku glaciers. Observations of the Taku Glacier were begun in 1890 when the U.S. Coast and Geodetic Survey mapped its terminus (map 8) and determined the bathymetry of the inlet. The terminus was presumably in fairly deep water then (fig. 109), but by 1941 the glacier was barely tidal. We established two stations on the north shore of the inlet in '26, both of which were overridden by ice about 1935. It turned out that the one closest to the glacier had been occupied in 1904 by Fred and Will Wright[16] and was last occupied in 1931 by Will, who referred to it as "Pond's Site," and by Wentworth and Ray (figs. 110–112).[17] I vividly remember seeing the Taku ice cliff (fig. 113). It was very spectacular in those early days. We took pictures but did not fix the position of the terminus. At Norris Glacier we visited what later came to be called Glory Lake. The trim line[18] of the high level of the lake surface when its outlet was dammed by Norris Glacier was still very distinct (fig. 114). You can barely see it now as the vegetation has grown up so much. We also saw the trim line of a recent advance, which we later determined to be about 1915. At Twin Glacier Lake we took pictures from a site that later became Sta. A, and visited with old Ben Bullard who had a farm in the Taku Valley (figs. 115, 116). We then returned to Juneau and continued to Glacier Bay.

Glacier Bay

For the trip to Glacier Bay, somebody had given us the name of a boat available for charter in Juneau, the M/V *Eurus.* The owner and skipper was Paul Kegel (fig. 117). He had an interesting history. As I recall, he was Hungarian, and had come over to this country some years before. He was a musician and had played the tuba in the New York Metropolitan Opera Orchestra for some years. He came to Alaska, probably in search of gold, which was the reason most people other than tourists came in those days. Paul was a lot of fun and good company. He'd been the jailer in Juneau for some years but had gotten sick of that job, so here he was taking us into the inlets of Glacier Bay and afterwards up the outer coast to Lituya Bay.

However, we soon found out that Paul didn't know very much about the boat and was not too good at reading charts. He knew how to start the boat and stop it and steer it a little bit, but the charts, such as they were, were something of an enigma to both him and the rest of us as well. He ran us aground on several sandbars that were shown on the charts (fig. 118), but we always floated off. At least once while at anchor, the boat swung into shallow water when the tide was running out. Tides are high there, sometimes over twenty feet between high and low. I remember that night while in Rendu Inlet, awakening to hear cans and other objects sliding across the floor of the cabin. We woke up to find that the boat was already well aground and listing. It went way over, but fortunately shipped no water before the tide floated it again.

One special objective of the trip was to visit Johns Hopkins Glacier which, as far as we knew, had not been photographed since 1912. Although several scientists had been to Glacier Bay since then, none had ever reported on the Johns Hopkins because of extremely heavy ice that filled the whole lower inlet. This ice prevented a boat from penetrating even as far as the big bend for a view into the upper inlet. Cooper's 1923 article in *Ecology* included a map of the Johns Hopkins[19] with the terminus the same as on the boundary survey map of 1912. I found out from him later that he never actually saw the terminus due to the amount of ice in the fiord. He just estimated its position. We found the same condition. Obviously all this ice was coming from Johns Hopkins Glacier, which was out of sight. Since we couldn't get the boat past the entrance of the inlet abreast of Lamplugh Glacier, Ben Wood, Andy Taylor and I climbed a ridge between the entrances of Tarr and Johns Hopkins inlets (map 11). At an altitude of around 2,500 to 3,000 feet we finally caught sight of the Johns Hopkins some six to seven miles beyond where it had been last seen in 1912 (fig. 119). The retreat of the Johns Hopkins was our principal

FIGURE 118

FIGURE 119

FIGURE 120

FIGURE 121

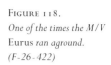

FIGURE 118.
One of the times the M/V Eurus ran aground. (F-26-422)

FIGURE 119.
The view up Johns Hopkins Inlet showing the amount of retreat of Johns Hopkins Glacier by 1926. It was poor weather and visibility wasn't very good, but the picture, although of poor quality, recorded what we saw. Aug 30, 1926. (F-26-248)

FIGURE 120.
Our camp at the junction of the third northern tributary and Geikie Glacier looking WNW up the tributary. Andy Taylor (L), Percy Pond, and B.S. Wood (R). (F-26-120)

FIGURE 121.
Andy Taylor at the boundary survey station GRANITE in Tarr Inlet, Aug 28, 1926. (F-26-226)

and most spectacular find that year, though we also recorded glacier-terminus positions in all the other inlets in Glacier Bay, as well as those visible at a distance from the boat along the coast while on our way to Lituya Bay.[20]

In Geikie Inlet we found Geikie Glacier receding from a prominent end moraine that had been formed across the valley, marking a small advance of the glacier about 1920. It actually had been in tidewater up until 1916. We packed camp five or six miles up the glacier (fig. 120), with the objective of going to the ridge dividing the Glacier Bay basin from the Brady Glacier so we could look west to view the Fairweathers [Range] from the east. Mount LaPerouse and Mount Crillon are the two southernmost peaks of the range, and we wanted to see if there was a route across the Brady Glacier and up the east side, which would provide access to those peaks. However, from the top of the ridge we quickly saw there was no practical route. It was a long way across the Brady and there was a very steep escarpment with hanging glaciers on the other [west] side.

Next we visited Hugh Miller Inlet. The glacier that had covered the inlet and reached Gilbert Island[21] up to 1921 had disappeared (map 12). Masses of stagnant ice were still on the banks of the inlet. At that time, you could go up the inlet, through a narrow channel past the northwest end of what at that time was Gilbert Island and out into the upper end of Glacier Bay. Gilbert Island was quite a massive island—it went up to about 1500 feet. It was the location of five survey stations of C.W. Wright in 1906, and of one of the boundary survey stations in 1907 (map 13).

We then continued up to Tarr Inlet (map 14). About 1913 Grand Pacific Glacier had retreated northward across the U.S./Canada boundary into Canada, making the head of Tarr Inlet a small Canadian harbor. We were pretty sure it was still well back behind the boundary. We did not know until after our trip that J.P. Forde, an engineer from Victoria, British Columbia, had visited the area in 1925 and mapped the terminus, showing it to be a little less than a mile behind the boundary. We took some pictures from the shore, and then Andy and I climbed to a bench about 3400 feet on the western slope of Mount Barnard, located on the eastern side of the fiord. We found a monument (fig. 121), obviously a boundary survey monument and, thinking it was the boundary, made some calculations as to where the Grand Pacific Glacier terminus was situated. Actually it turned out from the International Boundary Commission (IBC) maps issued in 1928 that it was Station GRANITE, which was at a distance of about 1200 feet within British Columbia, so our calculation of the location of the glacier terminus was considerably off. We

took pictures from the monument and it was interesting to compare the position of the terminus in those pictures with its position in pictures taken by the boundary survey in 1912 (fig. 122) when they established the station, and also in pictures taken by Lawrence Martin from down the inlet in 1911 (map 16).[22] The boundary survey checked the position of the boundary in '36. Jesse Hill, an engineer for the IBC who was in charge of that party, showed me some pictures of the boundary markers they had established on the east and west sides of the inlet. They became our Stations 1 and 2 in '41, but the glacier had advanced over them by the time of our return in '50.

One of the features we noticed on our trip in '26 was the movement down-glacier of a massive landslide on Ferris Glacier, the first northern tributary of Grand Pacific Glacier (fig. 123). I didn't figure out where the slide came from until 1950 on my way to Alaska, when I was examining the aerial photos taken by the U.S. Navy in 1948. In those days we took the train across the continent and a boat up the coast, offering time to study photos, read and write. In those photos I saw a great scar on the side of a mountain next to Ferris Glacier. It was a huge slide; half the mountain was gone. From pictures taken in 1912 by the Canadian Boundary Survey and by me in 1926, I calculated the annual movement of the glacier as it carried the slide on top of it between 1912 and 1926 at about 550 feet a year. I also determined the slide would have occurred about 1899, which would correspond to the September 1899 earthquake that shook that whole area.[23]

We went to Rendu Inlet next. Rendu Glacier was in the process of changing from a tidewater glacier to one terminating on land.[24] Actually, as happens in a number of places, the terminus moves very little or not at all, but the outwash builds up in front of the glacier, forming a bar. There was still a little of the ice cliff ending in water but the rest of the terminus was fronted by bars. The next time we saw the glacier in 1935, it was totally fronted by a bar. So when you repeat these observations, you see the processes that occur.

ے

Another advantage of photographic records is that a correct interpretation of the data is possible at a later date, regardless of the fact that the significance of the observation may have been overlooked at the time the picture was taken.

A. E. Harrison, 1954

We then went to Queen Inlet and occupied Triangle Island, a small rock island about a mile or so in front of Carroll Glacier (map 15). I found out later that this was one of Reid's stations in 1892. In pictures taken in 1919 by Mertie, the island was tidal—being under water at high tide. By 1926 I think a bar had formed in front of the glacier, so that high tide didn't even reach the island. Since then, Triangle Island has become completely surrounded by mud flats. In 1883, Captain James Carroll of the S.S. *Queen* sounded 96 fathoms between Triangle Island and the Carroll.[25]

From the Carroll, the next step was to go to Muir Inlet. The boat left Ben Wood and me at Camp Muir (near where John Muir had built his cabin in the 1890s) with a skiff, two tents and some provisions while it returned to Juneau for fuel. It was bad weather when we were dropped off, but the maps showed the glacier only a few miles from our camp (maps 17, 18). Paul Kegal said he thought it was farther than that and sure enough, when we could see the Muir terminus in the distance and time the arrival of the noise from the icefalls, we determined the distance to be about ten miles. We paddled up to the glacier, encountering a lot of ice on the way. We found sheltered water for the night in Goose Cove, which was about a half a mile from the glacier (fig. 124). We tied the boat up but during the

FIGURE 122A

FIGURE 122B

FIGURE 122.
Grand Pacific Glacier terminus from Sta. GRANITE in 1912 and 1926. (Upper photo by N.J. Ogilvie of the International Boundary Survey in 1912, No. D.M.D. 12. Margerie Glacier is in the upper left. Copy photo from the WOF Collection. Lower photo by W.O. Field, Aug 28, 1926, F-26-223.)

FIGURE 123.
The massive landslide on Ferris Glacier as seen from Sta. GRANITE in 1926. Andy Taylor is in the foreground. (F-26-221)

FIGURE 123

FIGURE 124.
*Ben Wood at our camp on
a shelf on the west side of
Goose Cove, Sep 2, 1926.
Goose Cove was completely
bare of vegetation because it
had been covered by ice only
six years before. We camped
on the moraine material,
which was still saturated
with water and awfully
squishy under the tent.
(F-26-264)*

FIGURE 125.
*Muir Glacier terminus from
my station (Sta. 3) above
camp at Goose Cove, Sep 3,
1926. The arrow points to
the Nunatak. (F-26-270)*

FIGURE 126.
*Muir Glacier terminus from
Sta. 3, Aug 29, 1958. The
arrow points to the Nunatak.
(F-58-K239)*

FIGURE 127.
*Ben Wood and the "To and
from the Glacier" sign we
found, a remnant of the days
when excursion steamers
landed passengers in this
area for a hike onto Muir
Glacier in the 1880s and
1890s. Sep 1, 1926.
(F-26-259)*

FIGURE 124

FIGURE 125

FIGURE 126

FIGURE 127

night some waves caused by an icefall off the Muir dumped the boat. The next morning the boat was there but upside down, hanging from the rope, with the oars gone. We used floor boards as paddles, but didn't look forward to the prospect of paddling ten miles through the ice back down the inlet to our campsite at Camp Muir. We decided to stay at the glacier another couple of days, even though it rained a lot of the time. We took a few pictures but conditions were poor (fig. 125). We occupied five stations and made a survey of the terminus (map 19). Our main station was right above where we camped, the highest knoll around, and I used that station for years.[26] The last time I was at that station was 1958, and the glacier was far up the inlet (map 20; fig. 126). The alders had grown up all around the station by then, making it difficult to reach. In 1926, there was not even grass or small plants around. We also visited an island that had just appeared from under the ice. Some pictures taken in 1919 by Mertie showed the glacier just over the island. We found a seal blind belonging to the local Indians, the Hoonahs, who apparently had been sealing there long before we arrived. I later proposed calling it Sealers Island.

We paddled back to Camp Muir with the floorboards after waiting for an outgoing tide and a favorable wind. While waiting at base camp for the *Eurus* to pick us up, we hiked around the moraines. And there we found the remnants of an old boardwalk and a sign that read, "To and from the glacier" (fig. 127). This had to date from the 1880s or 1890s, because no steamers had been there with tourists since 1899. Here, tourists, some probably in ordinary shoes, would be guided onto this boardwalk so they could walk to the ice without stepping on the wet, squashy ground. Now, the question arose whether to take the sign or leave it. I'm sure it would have been of interest to the museum in Juneau. Well, we made the wrong decision. We left it and by my next visit there in 1941, thick alders covered the whole area. I found out later that my mother and my grandparents had been up on that Muir Glacier boardwalk during a family trip to Glacier Bay in 1897 (fig. 128). When I found that out, I wished more than ever that I had taken the sign.

In the rest of Glacier Bay we found that Plateau Glacier had already retreated over a mile up what later became Wachussett Inlet, and that the Adams Glacier piedmont lobe still filled the valley of Adams Inlet.[27] Casement Glacier had retreated a few hundred yards from the shore of Muir Inlet, and Lamplugh and Reid Glaciers were near the entrance of their respective valleys. That ended the scientific part of the trip and we then headed north along the coast to Lituya Bay to hunt.

Lituya Bay

In '26 Lituya Bay, on the outer coast some fifty miles northwest of Cape Spencer, was not in Glacier Bay National Monument. In later years, it was acquired by the National Park Service and is now part of Glacier Bay National Park and Preserve. Lituya Bay is T-shaped and it has a very narrow, shallow and thus very tricky entrance at the coast. It is well known for this. It is probably an old moraine. The bay extends inland northeastward for ten miles. It's a large body of water, and at each tide, the water flows like a rushing river through the entrance, forming a real rapid. One has to go through the entrance in slack water, which is tricky. We went in, not exactly at slack water, and I remember looking over the side and seeing a great big rock that a large wave took us over. It was a dangerous place. That's where La Pérouse lost twenty-two men in a small boat in 1786.[28]

We anchored and went to look at the two glaciers at the head of the bay, the Lituya that comes in from the northwest and the North Crillon that comes in from the southeast. We took pictures of them and we set up a station on the beach (map 21), but it has never been reoccupied that I know of. The glaciers were known to be advancing, and at Lituya we did find great big stumps, three to four feet in diameter, where the trees had recently been knocked over by the advancing ice. Those have all been completely engulfed since then because of the glacier's further advance. It had been advancing slowly but evidently fairly steadily since La Pérouse's time. His map (map 22) is one of the earliest maps accurate enough to be reliable evidence of the positions of the termini. In those days the upper arms of the bay were each about two miles long to where the glaciers ended. Since then the glaciers have come forward so now each arm of the T is less than half a mile in length. This is interesting, because it shows totally opposite behavior from the glaciers on the east side of the Fairweather Range, in the Glacier Bay drainage, where we found all the glaciers either receding or fairly stable.

We climbed what was then known as Billy Goat Mountain. It has no official name. It is about 3,300 feet high and located on the west side of the upper end of the bay. We took pictures and it turns out that the site also had been the survey site of McArthur of the 1894 Canadian Boundary Survey,[29] so the photos taken in 1894 and 1926 show the changes during that thirty-three-year period (figs. 129, 130). Some pictures have been

FIGURE 128

FIGURE 129

FIGURE 130

FIGURE 128.
My mother (standing second from left) on a family trip to Muir Glacier in 1897. My Grandfather Sloane is standing at left, and my grandmother is standing third from left. (LaRouche photo, no number.)

FIGURE 129.
The view ESE to the head of Lituya Bay, Sep 9, 1926 from Sta. MCARTHUR. Cascade Glacier is in the lower left and North Crillon Glacier is in the right center. (F-26-325)

FIGURE 130.
Head of Lituya Bay in 1894 from Canadian Boundary Survey Station 213 (what I call Sta. MCARTHUR) on the mountain west of Gilbert Inlet. Lituya Glacier is in the lower left, Cascade Glacier in right center, and North Crillon Glacier is in the center right. (Photo by J.J. McArthur, #109A, of the Alaska Boundary Commission and filed as USGS Gilbert No. 1780; copy photo from the WOF Collection.)

FIGURE 131.
*Jim Huscroft, the hermit
who lived on Cenotaph
Island. Sep 11, 1926.
(F-26-343)*

FIGURE 132.
*Huscroft's home on
Cenotaph Island, Lituya
Bay. Sep 11, 1926.
(F-26-344)*

FIGURE 131

FIGURE 132

taken from other places in the bay, but as far as I know, we were the first ones to go back up there. It is a long, laborious climb.

While in Lituya Bay we met Jim Huscroft (fig. 131), the old hermit who lived on Cenotaph Island in nothing more than a shack, with a dock and a couple of outbuildings (fig. 132). Cenotaph Island is where La Pérouse buried the remains of the men who were drowned at the entrance. Huscroft had been there a number of years. I think he had a fox farm. He was known to row his skiff down the coast to Juneau once a year, pick up the newspapers from the previous year, bring them home, and read one every day—he was just one year late. He had quite a history.[30] I remember he came out to our boat and had tea or coffee with us. He showed us around Cenotaph Island, which is quite a good-sized island. I don't think anybody ever found the remains of the cenotaph for which it was named.

Meeting My Predecessors

Upon our return from Glacier Bay in 1926 we reported what we'd learned about the positions of the glacier termini to Lawrence Martin, chief of the Division of Maps at the Library of Congress in Washington, D.C. On the basis of our record, Martin had changes made in the International Boundary Commission map—which was in the final stages of completion—to indicate the extension of the inlets due to glacier recession at both Johns Hopkins and Muir. Martin had been to Alaska a number of times and he had a great influence on anybody who was interested in glaciers in Alaska. He certainly influenced me.

Martin then introduced me to Harry Fielding Reid, who had visited Glacier Bay in 1890 and 1892.[31] I remember that Martin and I took a train from Washington to Baltimore, and had dinner with Mr. and Mrs. Reid at their house. I was thrilled to meet Reid, who had retired as a professor at Johns Hopkins but still maintained a small office in the geology department there. He was not too old to go back to Glacier Bay, but had not done so. Naturally he was keen to know what had happened. After our initial visit, we saw each other from time to time and he did return to Glacier Bay with C.W. Wright from the USGS in 1931. I remember he always called me "Field." I last saw Reid just before I joined the U.S. Army in 1942. During the war he died at the age of 85.

All photographs of the end of a glacier are useful, especially those taken from a station easily accessible and easily described; photographs taken from the same station at a future date will show what changes have taken place in the interval.

H.F. Reid, 1896

When I returned from Glacier Bay to Cambridge [MA] in the fall of '26, the editor of *Appalachia*, the bulletin of the Appalachian Mountain Club, asked me to write an article on our trip. This appeared in the December 1926 issue. The title of the article was "The Fairweather Range: Mountaineering and Glacier Studies."[32] This was my first publication on glaciers. As I now realize, it contained little information about the glaciers, except it did have a map showing our routes and the major changes at Muir, Johns Hopkins, and Hugh Miller glaciers (map 23).

> *For the study of changes in the size of glaciers photographic views are of particular value. A view showing a glacier... in relation to details of adjacent land constitutes a record which can at any time be compared either with the objects themselves or with another photograph made in another year or month.*
>
> G.K. Gilbert, 1904

The 1926 trip also led to my meeting various people in the USGS and other agencies in Washington, who, with Lawrence Martin, were to influence and encourage this inexperienced but enthusiastic neophyte to carry on through a lifetime. Among these were Philip S. Smith,[33] then head of the Alaska Branch of the Survey, Fred Moffit,[34] J.B. Mertie, Jr., C.W. Wright, and the chief topographer Harvey Sargent.[35] And Stephen Capps[3,6] a big tall man with large steps I remember, who was sent to Alaska to look for minerals but always took pictures of glaciers. Also C. Hart Merriam[37] of the [USDA] Biological Survey, who had been on the Harriman Alaska Expedition of 1899.[38] They were all interested in what changes were occurring in the glaciers, but had to concern themselves with the primary mission of the Survey, which was hard rock geology, mapping, and locating mineral resources. I never met John Muir [died 1914] or George Frederick Wright, who were probably dead by then. These associations were very important to me, providing me with an unusual linkage with some of the earliest people who had been to Alaska, and giving me additional incentive to continue the observations and, in a very modest way, to try to follow in their footsteps.

Notes

1. Reid, "Glacier Bay and its Glaciers" (1896).

2. Cushing, "Notes on the Muir Glacier Region and its Geology" (1891, 207–230).

3. A biographical sketch of Grove Karl Gilbert can be found in Chapter 2, note 12. For a description of the Harriman Expedition in 1899 in which Gilbert took part and the resulting publications, see Chapter 2, note 8.

4. Lamplugh, "Notes on the Muir Glacier of Alaska" (1886, 299–301).

5. Wright, G.F., "The Muir Glacier" (1887, 1–18).

6. Vancouver, "Glacier Bay" (1798, 405–452).

7. See Appendix E for a brief summary of the Alaska-Canada boundary work.

8. Alfred Hulse Brooks (1871–1924) was a geologist who worked for the USGS engaged in Alaska explorations from 1894–1903. He served as chief Alaska geologist to USGS from 1903 until his death, and wrote a number of bulletins and professional papers. The Brooks Range is named for him.

9. Arthur F. Buddington (1890–1980) was a field petrologist and economic geologist. He was appointed assistant professor in geology at Princeton ca. 1920 and was chairman of the department from 1936–1950. Between 1912 and 1960 he spent forty-four summer seasons in the field, including five summer field seasons with the USGS in southeastern Alaska (*Geological Society of America Memorials*, 1984, 1–7).

10. William Skinner Cooper (1884–1978) was a world-renowned botanist and ecologist. He joined the botany staff of the University of Minnesota in 1915 and was appointed professor in 1929. His first scientific expedition to Glacier Bay was in 1916 to study plant succession in a newly deglaciated area. He also was intrigued with the fossil forests and realized then the importance of ascertaining the history of past glaciation and changes in climate. He returned to Glacier Bay in 1921, 1929, and 1935, and made brief visits in 1956 and 1966. He chaired a committee of the Ecology Society to explore the feasibility of designating the Glacier Bay area as a nationally protected area. In 1925, President Coolidge proclaimed Glacier Bay a national monument due to the efforts of that committee (*Memorial to William Skinner Cooper*, printed by the Geological Society of America, March 1980).

11. Cooper, "The Recent Ecological History of Glacier Bay, Alaska: I. The interglacial forests of Glacier Bay" (1923, 93–128).

12. John Beaver Mertie, Jr. (1888–1980) was a geologist who began working for the USGS in 1908. He became a full-time employee in 1911 and worked for the survey until his retirement in 1958 at age 70. He then

continued geologic investigations on a part-time basis until 1975. He was considered a pioneering scientist for his contributions to the regional and economic geology of Alaska. His last publication in 1979 was on the hundredth anniversary of the Survey and in the seventy-first year of his association with that organization (*Geological Society of America Memorials*, 1982).

13. See Carpé, "The Mt. Logan Adventure" (1933); Lampart, "Conquest of Mt. Logan, 1925" (1926).

14. See Carpé, "The Conquest of Mt. Fairweather" (1931); Ladd, "The Fairweather Climb" (1932); Moore, "Mt. Fairweather is Conquered at Last" (1931).

15. Lloyd Winter and Percy Pond opened a photo shop in Juneau in 1893. For fifty years their work captured the daily life and natural splendor of the Juneau area. Takuakhl, their cabin located up the Taku River with a scenic glacier view, was a popular retreat for them (Alaska Historical Library, Juneau, Alaska).

16. Charles Will Wright (1879–1968) was chief of the mining division of the U.S. Bureau of Mines (1928–1935) and chief foreign mineral specialist for the U.S. State Department. He joined the USGS in 1903 and was engaged in mine reporting and in mapping the geology of southeastern Alaska until 1909. He and his brother F.E. Wright visited Glacier Bay in 1906 for the survey to study geology and glacial geology. Will returned to Glacier Bay, again for the USGS, with H.F. Reid in 1931 (W.O. Field, personal communication. See also note 33, this chapter).

Frederick Eugene Wright, 1877–1953, was an assistant geologist for the USGS during the summer of 1904 and 1905, and served as geologist from 1906–1917. While with the survey, he made a special study of the geology of southeastern Alaska. From 1906 until his retirement in 1944 he was staff petrologist in the Geophysical Laboratory of the Carnegie Institute of Washington, D.C. (*National Cyclopaeda of American Biography* 41).

17. Chester Keeler Wentworth, 1891–1969, was a volcanologist, geomorphologist, and hydrogeologist. He is most widely known for the Wentworth scale for classification of clastic sedimentary rocks. He was appointed associate professor of geology at Washington University in 1928 and spent the summer of 1931 traveling down the Yukon River studying glaciers and glacial deposits. From 1933 until his retirement in 1951, he served as geologist for the Honolulu Board of Water Supply (*Geological Society of America Memorials*, 1973, 97–104).

Louis Lamy Ray (1909–1975) a geologist, went to Circle City, Alaska in 1931 as an assistant to C. K. Wentworth. He proposed the preparation of a gazetteer of glaciers, an idea that was favorably received by François Matthes (see Chapter 6, note 23). Wentworth and Ray gave papers on their Alaska work at the 1932 and 1934 annual meetings of the Geological Society of America, and in 1936 they published "Studies of Certain Alaska Glaciers in 1931," a significant paper on Alaska glaciers (*Geological Society of America Memorials*, 1980, 1–8. See also Chapter 6, note 16.)

18. A *trim line* is a very sharp boundary line that marks the maximum lateral extent of a glacier. The line may be the limit of erosion in bedrock or modification of vegetation by a glacier.

19. See note 11 above, figure 1, for Cooper's map of the Johns Hopkins.

20. See Appendix C for a trip summary.

21. Gilbert Island is no longer an island due, in part, to isostatic rebound. When additional weight, in this case glaciers, is placed on an area of the earth's crust, the land subsides under the load. If that weight is removed, such as by the melting of the glaciers, the area will rise, or "rebound." In the case of Glacier Bay, "the greatest change in shoreline of any place in the United States during historical times" has occurred. When George Vancouver visited Glacier Bay in 1794, he found the bay almost filled with ice, and by 1892 most of the retreat had taken place (Shepard and Wanless, *Our Changing Coastlines*, 1971, 405).

22. Colonel Lawrence Martin (1880–1955) a geographer and glaciologist, was a member of that distinguished group of builders of modern American geography whose early professional training was largely geological. He was contributing editor of the *Geographical Review* from 1922. He became a lieutenant colonel in the Officers' Reserve Corps after World War I and then served from 1924 until his retirement in 1946 as chief of the Division of Maps in the Library of Congress. Martin's glaciological studies began in 1904, when he went to Alaska

as a member of a U.S. Geological Survey party. He returned to Alaska the next year with a grant from the American Geographical Society, and together with Ralph S. Tarr began a long series of studies of glaciers and their variations. Later expeditions were sponsored by the National Geographic Society, under the leadership of Tarr and Martin in 1909 and 1911, and of Martin alone in 1910. In 1913, Martin served as a guide for the excursion of the International Geological Congress up the coast of southeastern Alaska to Yakutat Bay. After the excursion, he and Eugene Romer, a well-known Polish geographer, visited Glacier Bay. Tarr and Martin, jointly and individually, published a large number of papers on the glaciers and glacial geology of coastal Alaska. Their book *Alaskan Glacier Studies* (chapter 2, note 13) is still considered their monumental work and is one of the finest volumes on glaciers to appear in this country (*Geographical Review*, 1955).

23. Tarr and Martin, *The Earthquakes at Yakutat Bay, Alaska, in September 1899* (1912).

24. A *tidewater glacier* is one that terminates in salt water.

25. 96 fathoms = 176 meters = 576 feet.

26. "That station" refers to Muir Sta. 3.

27. A *piedmont lobe* forms when a valley glacier extends out over a plain and spreads into a great lobe-shaped pattern.

28. *A voyage round the world performed in the years 1785, 1786, 1787, and 1788 by the* Boussole *and* Astrolabe *under the command of J.F.G. de La Perouse* (1799).

29. Sta. 13 was established in 1894 by J.J. McArthur of the Alaska Boundary Commission. W.O. Field refered to it as Sta. MCARTHUR "in honor of that fine Canadian surveyor" (personal communication). In Bill's collection, there are many references to the "Canadian Boundary Survey" when the pre-1900 boundary survey work is mentioned. The group responsible for the initial survey of the Alaska boundary, 1893–1895, in fact was called the Alaska Boundary Commission and was a joint enterprise between the United States and Canada. See Appendix E for a description of the boundary survey work.

30. Bohn and Brower, *Glacier Bay: The Land and the Silence* (1967, 31–32).

31. Harry Fielding Reid (1859–1944) was a professor of dynamic geology and geography at Johns Hopkins, a position that he held from 1911 until his retirement in 1930 as professor emeritus. He worked as a special expert in earthquake records for the U.S. Geological Survey from 1902 to 1914. He was a member of the Commission International des Glaciers and author of numerous reports and articles on glaciers, earthquakes and similar subjects. He studied Muir Glacier in 1890 and extended his observations to the whole of Glacier Bay in 1892. He returned with Charles Will Wright (note 16 above) in 1931. Reid Glacier in Glacier Bay is named for him ("Memorial to Harry Fielding Reid," 1944, 295–298).

32. Field, W.O., "The Fairweather Range: Mountaineering and Glacier Studies" (1926).

33. Philip Sidney Smith 1(878–1949) was chief of the Alaska Branch of the USGS. He taught geology and physiography at Harvard from 1900 to 1906, and joined the Survey in 1906 as an assistant geologist, retiring in 1946 (*New York Times*, Obituary, May 11, 1949).

34. Fred Howard Moffit (1874–1958) worked for the USGS in Alaska for forty years and authored over fifty publications on Alaska geology and mining. Mount Moffit, eleven miles southeast of Mount Hayes in the Alaska Range, is named for him (Orth, *Dictionary of Alaska Place Names*, 1971).

35. Rufus Harvey Sargent (1875–1951) worked for the U.S. Geological Survey as a topographer who did extensive exploration and mapping on the Kenai Peninsula, Alaska. The Sargent Icefield on the Kenai Peninsula is named for him (Orth, *Dictionary of Alaska Place Names*, 1971).

36. Stephen Reid Capps (1881–1949) joined the USGS in 1908 and did aerial and economic studies in Alaska until 1936. He was appointed assistant chief geologist of the USGS in 1942. He was especially interested in the scenic wonders and game resources of the country adjacent to Mount McKinley in interior Alaska and was one of the sponsors in the movement to set aside that area as a national park. Mount Capps in Denali National Park and Preserve and Capps Glacier in the Tordirillo Mountains in Alaska are named for him.

37. Clinton Hart Merriam (1855–1942) practiced medicine from 1879 to 1885 when he became chief of the U.S. Department of Agriculture Biological Survey, a post he held from 1885 to 1910. He was a member of the Harriman Alaska Expedition of 1899 and was editor of the *Harriman Alaska* series, the thirteen-volume series describing that expedition (see Chapter 2, note 8 for a brief description of the voyage and the resulting publication). He was founder of the National Geographical Society and chairman of the U.S. Geographic Board from 1917 to 1925. Mount Merriam in Glacier Bay National Park and Preserve is named for him.

38. See Chapter 2, note 8 for a description of the Harriman Alaska Expedition of 1899 and the resulting publications.

FIGURE 133

CHAPTER 4

A FAMILY EXPEDITION
1927

In 1927 there was one last nonscientific, family trip to Alaska. Papa wanted to go to Alaska—he had heard about it from me. So in August the whole family boarded the S.S. *Yukon* in Seattle and we went up the coast (fig. 133). E. Lester Jones, U.S. Commissioner for the U.S.-Canada boundary survey, just happened to be along on the journey. We enjoyed talking with him. My mother, sisters, and brother returned to Seattle on the same voyage. My father, my father's friend Dr. Bruce Paddock from Pittsfield [MA], Rocky Bonsal, one of the friends with me on my '26 trip to Glacier Bay, and I got off at Cordova, and went by the Copper River and Northwestern Railroad up the Copper River to Chitina (figs. 134–137). There we began a three-day trip over the Richardson Highway to Fairbanks (map 24). On the way to Fairbanks, we stopped at the various roadhouses, some of which are still there[1] (figs. 138–143). From Fairbanks we traveled on to McKinley National Park where we organized for the remainder of the trip (figs. 144–146). From

McKinley National Park we went by pack train down along the north side of the Alaska Range, past Mount McKinley and Mount Foraker to the west end of the park towards Tonzona Basin to do a little hunting for sheep (map 24). The trip was organized by a promoter in Fairbanks. A man by the name of George Lingo was in charge of the trip, even though he was the youngest man in the group. The oldest men in the group, the guides Bob Ellis and Van Bibben, were in their seventies. You could only drive about twenty miles on the road from McKinley Park and then you had to transfer to horses, so you had a long way to ride down the range.

We got to Wonder Lake for views of Mount McKinley, and we stayed there with John and Paula Anderson. They had a fox farm overlooking Wonder Lake (fig. 147). Paula was quite a lady—a short, funny, stylish lady. She told of how one winter she was sick and John decided they better get her to Anchorage to see a doctor. It was fifty miles or so by dogsled to where they would catch

FIGURE 133.
The Yukon *in port in Ketchikan.*
(WF-27-F563)

45

FIGURE 134.
The train station at Cordova. (WF-27-F613)

FIGURE 135.
The Copper River and Northwestern Railroad. (WF-27-F502)

FIGURE 136.
While in Cordova, we went out to Mile 49 again but instead of the train, we went by what was called the "speeder." It was a Model A Ford with flanged wheels that ran on the tracks, and was a very convenient way of getting up and back over the 49 miles of track. (F-27-F54)

FIGURE 137.
The Hotel Chitina. (WF-27-F505)

FIGURE 138.
Copper Center. (F-27-F51)

FIGURE 139.
The Sourdough Road House on the trip to Fairbanks. (F-27-F61)

FIGURE 140.
The Gulkana Trading Post. (F-27-F50)

FIGURE 134

FIGURE 136

FIGURE 135

FIGURE 137

FIGURE 138

FIGURE 139

FIGURE 140

FIGURE 141

FIGURE 142

FIGURE 143

FIGURE 144

FIGURE 146

FIGURE 145

FIGURE 147

FIGURE 141.
The Richardson Highway.
(WF-27-F632)

FIGURE 142.
Crossing the Tanana.
(WF-27-F635)

FIGURE 143.
The Chatanika Post Office/
Trading Company and the
Nirich Grill in 1927.
(WF-27-F641)

FIGURE 144.
The park service facilities at
McKinley National Park in
1927. (WF-27-F651)

FIGURE 145.
This is the park headquarters
at McKinley. The lodge
hadn't been built yet. (WF-
27-F654)

FIGURE 146.
Harry Karstens (L) was the
superintendent of the park.
He made the first ascent of
Mount McKinley (South
Peak) with Hudson Stuck
in 1913. Karstens Ridge is
named after him. That is one
of his rangers on the right,
perhaps Grand Pearon.
(WF-27-F657)

FIGURE 147.
John and Paula Andersons'
home at Wonder Lake.
(F-27-F94)

FIGURE 148.
Here I am with Hiu
Skookum. (F-27-F363)

FIGURE 149.
The obituary (A) for Hiu
Skookum (B) in the 1938
volume of Appalachia.
(Printed with permission.)

FIGURE 150.
The packtrain leaving
Tonzona Valley on the
return trip. Mount
McKinley is in the distance.
(F-27-F208)

FIGURE 148

HUTS AND TRAILS

Hiu-Skookum, 1927–1938.—Pinkham Notch Camp has lost a faithful old friend, Hiu-Skookum, who passed peacefully away in his eleventh year on August 29, 1938.

"Hiu" was one of a litter of four puppies born in Mount McKinley National Park in Alaska and brought out from there in 1927. William Osgood Field, the famous globe-trotter and glaciologist offered him to me in April 1930, and ever since then he has been a popular member of our Pinkham Notch Camp staff. "Hiu stood for the finer and higher things in mountain hospitality" writes Jim Taylor, "and whether it was for a friendly romp up the mountain, or a singing and yodeling session in the cook shack, he was always present."

A gentle and stately old fellow, he was much loved by thousands of people, but he was an independent dog and did not care much for the usual palaver or petting that some people felt he needed. Many times he would go off by himself, or with some of the hut fellows, to remote points in the mountains, and he knew the Mount Washington Range as well as any human. There is record of his having been taken to Greenleaf by one of the hut boys, but becoming uneasy there, he struck out for home on his own. He passed Galehead, Zealand and Lakes-of-the-Clouds Huts, on his way, climbed over the summit of Mount Washington, and having apparently travelled the entire distance by trail, turned up at Pinkham Notch Camp just sixteen hours after he had been missed at Greenleaf. He also returned numerous times by himself from Madison and Carter Notch Huts.

"Hiu" had no pedigree, other than being a cross between a malemute and a shepherd of some kind, and true to his Alaskan upbringing he slept out in the snow many nights in sub-zero temperatures; covering his nose with his tail, and lying curled up in a big fuzzy ball. The day after he died I found him in his old nook in the woods, where it had been his custom to lie on the bank of the river in the sun.

Hiu-Skookum is the Alaskan Indian word for "Very Fine," and Hiu-Skookum of Pinkham Notch Camp lived up to his name to the very last. We all miss him a great deal.

JOSEPH B. DODGE,
Pinkham Notch Camp

FIGURE 149A

team. The doctor in Anchorage asked where they came from and then asked Paula if she had had any exercise lately. "I think lack of exercise is the problem with you," he said.

The Andersons had some pups when we were there. We bought two of them and took them with us. I called mine Hiu Skookum, which is Eskimo [Chinook] for "very good" (fig. 148). In Seattle on the return home I had a box full of film that was not packed very well. The railroad said they would take the dog but that the box needed to be repacked. Now, I had to make a decision—go with the dog or stay and pack the film. I went with Hiu, and a friend of mine in Seattle, Joshua Green, repacked my film and shipped it along later. Hiu ended up on my family's dairy farm in Lenox. He was a wonderful dog, sweet and gentle, but also the suspect twice in a break into the sheep pasture with unfortunate results for the sheep. I called my friend Joe Dodge, who ran the Appalachian Mountain Club cabin at Pinkham Notch, and told him that I had a wonderful dog who had a liking for sheep and because he had no sheep, would he take the dog. So Hiu went to live at Pinkham Notch. Everybody knew him and when he died years later there was a picture and obituary for him in *Appalachia*, the journal of the Appalachian Mountain Club (fig. 149).[2]

At Wonder Lake, my father's horse shied and jammed my father's leg against a rock outcrop and broke his ankle. It was so swollen that the doctor wouldn't allow him to go any farther. They went back to Seward and eventually returned home. Rocky Bonsal and I continued on to complete the objective of getting down to the Tonzona basin on the west side of McKinley National Park to do a little more hunting. I photographed several glaciers down there. One at the head of Tonzona Basin is now called Surprise Glacier. I don't think I've ever seen a picture of it taken since then. It would be interesting to see how it looks now, more than sixty years later. We passed very near the

FIGURE 149B

FIGURE 150

FIGURE 151

FIGURE 151.
Russ Merrill's plane
Anchorage No. 1. *From a
postcard I bought in Alaska
in 1927.*

FIGURE 152.
*The boat that we took out of
Kasilof. (F-27-F490)*

big glaciers that come off the Alaska Range—the Muldrow, the Peters, the East and West Foraker (fig. 150). The East Foraker Glacier now is called the Straightaway Glacier. All of those glaciers are so different from what I'd seen on the coast. They end in high moraine piles and in some cases it is very hard to see where the glacier ends and where the frontal moraine begins. I have never been back there.

On the way back to Anchorage, the train stopped for the night at Curry where there was an inn for passengers. While there I received a phone call from a man I'd met earlier who ran an outfitting agency in Anchorage. He said there was an opening at a camp down on Tustumena Lake on the Kenai Peninsula for moose hunting, and asked if I would like to change my plans and spend ten days down there. He said somebody would fly me down. So in due course Russ Merrill,[3] the flyer for whom Merrill Field in Anchorage is named, flew me out—my first-ever plane flight (fig. 151). We stayed at Moose Camp, from where we did a little moose hunting. By then, however, I was becoming more interested in taking pictures than in shooting animals.

We came back to Anchorage by boat out of Kasilof on Cook Inlet (map 24). This boat ran back and forth from Kasilof to Anchorage (fig. 152). There was no other way to get to Anchorage—everything was by boat. We boarded the boat in the afternoon and hit rough water. Waves were breaking over the pilothouse. Everything was tightened down, all the doors and windows closed, the gasoline engine in the middle of the cabin where we ate was going "Putt-putt-bup-bup," and the exhaust fumes were terrible. Between that and the seasickness, I really was ill. Two of us didn't want to live through the night. We made it to Anchorage the next morning and climbed out, glad to be still alive. This was my last hunting trip to Alaska. All my future trips to Alaska were to photograph and survey glaciers.

FIGURE 152

Notes

1. See Appendix E for a description of the roadhouses pictured in the book.

2. *Appalachia* (1938, 285, 287).

3. Russell Hyde Merrill was one of the pioneers of aviation in Alaska. He has been credited with many firsts: first successful attempt to fly a single-engine plane across the Gulf of Alaska; first commercial air flight westward from Juneau; discoverer of the pass that opened a new and shorter air route to the lower Kuskokwim on November 8, 1927, now known as "Merrill Pass"; first to make a night landing in Anchorage on November 24, 1927, upon returning from Ninilchik with a critically wounded school teacher. He was last seen on September 16, 1929, as he took off for Sleetmute and Bethel. (From an information sheet obtained by the editor at the airfield and news clippings in the WOF Collection.)

FIGURE 153

CHAPTER 5

FILMS AND TRAVEL
1927–1935

The University Film Foundation

Although I was becoming more and more
interested in photographing glaciers, my main
occupation from 1927 to 1935 was filmmaking
and world travel, two other favorite activities.[1]
Ever since I could remember, my father was
interested in taking pictures. Besides the pictures
he took in the 1890s during his inspection trips
along the New York Central Railroad to illustrate
his reports, he also recorded what went on in our
family life, as well as people who visited us. He
had a movie camera too, and took moving pictures
beginning in the first decade of this century. So,
from the very beginning I was exposed to moving
pictures as well as still photography. I had a
Brownie camera by the time I was ten years old,
and I think I began taking pictures as soon as I
could hold the camera steady.

My interest in photography and movie making
led to my involvement with the University Film

Foundation at Harvard. Its function was to assist
in the showing of films in classrooms by providing
a projector and a projectionist, a screen, etcetera.
I was not a paid employee. Some were paid, but
it was a struggling, nonprofit organization. The
director was Jack Haeseler, who had graduated
a few years earlier than I. The foundation also
made educational films. Haeseler had made a
documentary film of the life of the Bedouins, a
tribe in North Africa. This was the kind of film
that we see quite frequently on TV nowadays, but
in those days, there wasn't much opening for it.
Haeseler was a pretty good cameraman and he
taught me a lot of filming techniques. However,
as sound gradually came in during the early '30s,
it tripled the cost of producing this type of film,
making them prohibitive for us to produce.

The first travel picture that I recall making
was in Africa in 1928. William N. Beach, an older
man that I knew, took me along on a trip to Africa
where he wanted to hunt. He obtained museum

FIGURE 153.
*Our wedding picture from
the* New York Times.
(April 5, 1929, p. 18)

FIGURE 154.
My enthusiastic guide,
Muzahan Dadeshkiliani.
Oct 30, 1929.
(F-29-C209)

FIGURE 155.
The view near the top
of Dongus-Orun Pass,
crossing the backbone of the
Caucasus. Swanetia is on
the other side. We had to
cross a glacier to reach the
pass. I was here again in
1971, but couldn't obtain
permission to walk over the
pass into Swanetia. Aug 14,
1929. (F-29-C7)

FIGURE 156.
On the first trip, we
continued down the other
side into the densely
forested Nakra River
valley and began to see
the Swans, the people of
Swanetia. The village of
Tavrar (or Tavrari) was the
first settlement we reached.
Tavrar was in what was
called Princely Swanetia.
The princes held onto power
here until the end of World
War I and then the area
was taken over by the Red
Army in 1921. This man
is a member of the Tavrari
Ispolkom, the executive
council for the village or
district, and he is shown
with his family. Aug 17,
1929. (F-29-C26)

FIGURE 154

FIGURE 155

FIGURE 156

support by arranging for Bill Brown, chief taxidermist for the U.S. National Museum—part of the Smithsonian Institution in Washington, D.C.—to accompany us. He was to obtain specimens to replace some of the animal capes that were disintegrating.[2] I was along to take pictures and to assist him. Bill Brown and I boarded the S.S. *George Washington* in mid-January. His wife saw us off and the last thing she said to him was, "Don't try to be a hero, come back alive." He'd never been further than Detroit and here he was heading for Central Africa in 1928!

We traveled up the Nile in a chartered boat and saw all the sights along the river, both the animals and the human beings that lived along the bank—the Dinkas, the Shilluks, the Nuers. This was in the winter, the coolest and driest time of the year there. During the rest of the year there is more rain and the people go back to farm some miles from the river, but in February, March and April they tend to live beside the Nile because the interior is very dry. I edited the footage we took and made a film that I lectured with a few times under the title *Up the Nile to Central Africa*. That trip gave me some experience taking pictures under not very good conditions, using a very big, heavy tripod, and a heavy, awkward camera. I had a big French camera, a Debrie, and I had the same problems panning and following moving animals through the bush or across the plains as Carl Akeley found about the same time. He later designed a camera, named after him, that had a gyroscope tripod so one could film smoothly up and down or sideways or both, instead of a separate horizontal and vertical crank.

On that trip I met my first wife, Alice Withrow. At dinner on the ship on the way over, I was seated at a big round table hosted by the chief engineer. There must have been ten people at least, including an interesting-looking young lady across the table. During dinner, I mentioned that my brother was at the London School of Economics. The young lady said that she was on her way there. This set up a conversation, and we saw quite a bit of each other on that trip across the Atlantic to Cherbourg, France. It was a long trip, a good nine days to get across, and the old *George Washington*, a coal burner, rolled all the way.

Alice was different from anybody I'd met before. Her mother had been a school principal, and her father was a professor of chemical engineering at the Ohio State University. She was only about eighteen, bright and intelligent, and she wanted a change. She had been to Ohio State for a year or two, and she had jumped at this chance to gain some experience in Europe, which is now common but wasn't in those days. So she was sort of fleeing from home. She disembarked

FIGURE 157

FIGURE 157.
The Swans were hard working farmers, happy, very cordial, very proud and very pleased anybody would come to see them. These people are from Tavrari. Aug 17, 1929. (F-29-C23, 24)

at Southhampton, England, and Bill Brown and I left the ship at Cherbourg. We continued our trip down through the Mediterranean and the Suez Canal to the Sudan with Mr. and Mrs. Beach. When I returned in the spring, I went to London to see Alice. To make a long story short, Alice and I were married a year later. Her father had said she could not marry until she was twenty. She knew he couldn't remember her birthday, so as a last trick to pull on her father, she married the day before she was twenty (fig. 153).

The Hidden Valleys of the Caucasus

After Alice and I were married in '29, I made my first trip to the Caucasus in Georgia (map 25). I knew the people at the Open Road, a travel organization that arranged trips to unusual and interesting places, and they asked me if I would be interested in joining an American group for a twelve-day trek across the Caucasus, passing through a place called Swanetia. Geographically, the Caucasus are the limits between Europe and Asia, but Georgia to the south in Transcaucasia is more European than it is Asian. I had already read about the Caucasus because there were high, magnificent mountains there.[3] The early mountaineers in the Caucasus were mostly British—such as Douglas Freshfield, an eminent geographer—so I knew from his descriptions that it was beautiful, fascinating country, but I didn't know much else about it or the interesting people who lived there.[4]

The main source of interest, as it turns out, *is* the people. The whole area has been called an ethnological museum. One of the most interesting groups of people in all the Caucasus is the Swans, the inhabitants of the little district of Swanetia on the southern slope of the Central Caucasus. The high valleys where they live are in the heart of the main range, in the immediate neighborhood of some of the highest and most magnificent peaks, such as Ushba, Tetnuld, Elbruza and Shkara. In many cases these recently glaciated valleys have been cut off by a gorge, and the people are then able to protect themselves from outsiders along these gorges at the entrances to the valley. They grew all the necessities of life except salt and kerosene. Many of these people had moved there during times of turmoil, which were almost incessant in the history of the Caucasus. There was always a strong tendency for fighting between the Turks in the south and the tribal groups that preceded the Russians in the north, and they usually clashed in the Caucasus. Books by both Baddeley and Allen document the long history of the Russian Empire's fight for this area.[5]

I took still pictures and about 1200 feet of 35-mm film on the twelve-day trek. I showed the film a few times when I returned. I said that I'd like to go back and finish this job of traveling through this area taking pictures. It was nonpolitical, scenic, and interesting in an ethnological way. I thought that if Americans saw it, it could make friends. I also was fascinated by the possibility of documenting a changing society, an isolated one emerging from a way of life little changed in hundreds of years, a society coming out of the twelfth into the twentieth century. So

FIGURE 158.
When the Bolsheviks first
sent in military units during
World War I, they were met by
determined farmers who would
not let them up the Lenheri
Gorge in the Ingur Valley. This
is Prohoroff's Rock in the gorge
(also known as Red Officer's
Rock) named after the captain
of that unsuccessful Red Army
unit. It took four years before
the Reds were able to get in
there and take over. That was
in 1924. In 1929 this was a
place still in transition. Here
the trail is cut out of solid rock.
Laraquaqua is the local name
for the place. (F-29-C84)

FIGURE 159.
The farmers were simple and
in general very pleasant and
hospitable. This is an old
Swan at Deis. Oct 24, 1929.
(F-29-C91)

FIGURE 160.
The primary school and the
children in Lachamul. The
people in the village were
proud that the school included
girls. Oct 25, 1929. (F-29-C98)

FIGURE 161.
Some of the people live to a
good old age in this area.
These children are with their
grandmother. Oct 29, 1929.
(F-29-C100)

FIGURE 162.
While in Becho we stayed
with German (pronounced
Gair-mahn'). He was a priest,
schoolteacher and blacksmith
as well as a very jovial man
and a lot of fun. Oct 30,
1929. (F-29-C218)

FIGURE 163.
Our interpreter Leo
Kondratov, with German's
sword. Leo's English was not
very good, but we got along
and he was a really a great
help. (F-29-C219)

FIGURE 164.
(Opposite) German's house
with Ushba in the distance,
not the highest peak in the
area but a very spectacular
(continued)

FIGURE 158

FIGURE 160

FIGURE 161

FIGURE 163

FIGURE 159

FIGURE 162

FIGURE 164

at a cocktail gathering in New York one evening, a man who had been sitting at a desk in the corner handed me a piece of paper and said, "Here's your permission to go back." It turned out that he was the head of the local political hierarchy in Swanetia. So I returned there for most of two months that fall and was able to travel anywhere I wanted (map 26) to take photographs and to finish the motion picture coverage begun in the summer.[6] I also wrote an article[7] dealing with the mountaineering aspects of the area.

My enthusiastic guide was Muzahan Dadesh-0kiliani, best described as a magnificent sort of man (fig. 154). He had been a prince whose family had controlled the lower part of this area up until the [Bolshevik] revolution.[8] He was in the Russian Army on the western front during World War I, taken prisoner by the Germans, and finally released. Now he was quite happy as a simple member of the executive council. His job was to organize singing and dancing and festivities, a job for which he was well qualified. He lived in Mestia, the capitol of Swanetia, where he had been given an apartment and a horse. I remember his telling me he no longer owned any land, but at least he had something to do. He knew everyone and was well liked by the farmers. He shepherded us around and people were pleased to see him. They were honored at the visit by a foreigner who was interested in spending time with them, photographing their farms and their way of life and taking them seriously (figs. 155–187). The most difficult part of the trip was coping with the welcoming festivities. Every evening the villagers had to pull out the home brew and see that their guests were properly taken care of. Some of those evenings I just went to sleep in a corner while Muzahan partied. The last night in Mestia we had a party in which all sorts of people showed up, including the chief of police. This was a men's party so the big event was to drink people under the table. I was fairly careful, but I recall that the chief of police ended up passed out on the floor.

When I returned home, Amkino wouldn't allow the film I made to be distributed commercially, so I lectured for several years with it (fig. 188).[9] It went fairly well, but I don't like to lecture. I steered away from the political aspect of whether communism was good or bad, and just made a plea for friendship and understanding. I wasn't as good at that as Julien Bryan.[10] He balanced the pro/anti-Communist issue very well. As far as I could make out, most people in the Caucasus had no political affiliation one way or another; they just wanted to be left alone. They didn't mind a little progress here and there, if it didn't interfere with their way of life too drastically.

FIGURE 165

FIGURE 166

FIGURE 167

FIGURE 168

(Figure 164 continued)
German lived in what was known as a European-type house, meaning it had windows and it was made of wood instead of being a walled enclave like most of the people lived in. There was a chimney so the fire wasn't in the middle of the dirt floor, and his cattle were in a separate building rather than in the basement. It was more comfortable than most of the houses. One night at this priest's home while I, an American, was there with Muzahan, an ex-prince, two members of the Communist Party labor union came through and were guests for the night. That was the kind of combination that couldn't have happened anywhere except in the more remote and somewhat less rigid areas of the Soviet Union. In Moscow and most parts of the USSR such a group would not have had any contact with each other. It was a most unusual combination but everyone got along well. Oct 29, 1929. (F-29-C105)

FIGURE 165.
The teachers and school children in Becho. Oct 30, 1929. (F-29-C124)

FIGURE 166.
A cooperative store in Becho. Oct 30, 1929. (F-29-C121)

FIGURE 167.
The high school and clinic in Becho. Oct 30, 1929. (F-29-C122)

FIGURE 168.
The main square in Tsomar where Muzahan lived before the Bolsheviks took over. Oct 31, 1929. (F-29-C129)

FIGURE 169.
Muzahan's former house and tower in Tsomar, which was taken from him during the revolution—as a prince he hadn't lived in any palace! It was just a little better than the local farmers' houses. Oct 31, 1929. (F-29-C128)

FIGURE 170.
Our outfit in the courtyard of the schoolhouse in Kali. It was the kind of schoolhouse in which we frequently stayed. Nov 6, 1929. (F-29-C183)

FIGURE 171.
The Kali school teacher (center). Nov 6, 1929. (F-29-C182)

FIGURE 172.
The monastery of Kirrik and Eflit, one to two miles from the village of Kali. Aug 20, 1929. (F-29-C56)

FIGURE 173.
These two guards of the monastery of Kirrik and Eflit met all strangers with suspicion and ready rifles. Aug 20, 1929. (F-29-C59)

FIGURE 169

FIGURE 171

FIGURE 170

FIGURE 172

FIGURE 173

FIGURE 174

FIGURE 175

FIGURE 176

FIGURE 177

FIGURE 174.
The uppermost valley of the Ingur River leading to the village of Ushgul was a very remote area where few foreigners had been. There were several different communities. The people here worked all year round. In many locations, the men had become the warriors and tended to just stand around with spears looking impressive, but here everybody had to work because life was extremely hard. This is the lower village of Ushgul. A very prominent part of each village was the siege towers. These towers in Swanetia date back to the fourteenth or fifteenth century and were built for defense against foreign attacks as well as protection during the devastating blood feuds that at times were rampant in many of these communities. Almost anything you did in courtship would antagonize somebody. The siege towers were very much in use until the Bolsheviks came in and helped restore order. Nov 3/5, 1929. (F-29-C168)

FIGURE 175.
Upper Ushgul from the Church of the Virgin (see figure 183). Nov 3/5, 1929. (F-29-C173)

FIGURE 176.
A woman spinning wool in Gabiany, Ushgul. Nov 3/5, 1929. (F-29-C172)

FIGURE 177.
We went as high as we could, into the snow on a ridge above Ushgul, to get pictures. I had an Akeley camera. Two men from Ushgul are with me. The ridge of Shkara is in the lower left. Nov 4, 1929. (F-29-C155)

FIGURE 178.
From this ridge above Ushgul you could see Shkara (17,038 feet), the third highest peak in the Caucasus and one of the most formidable for mountaineers. Nov 4, 1929. (F-29-C157)

FIGURE 179.
Tetnuld (15,918 feet) and the Adish Glacier. There is quite a history 3 of attempts to climb these high peaks in the central Caucasus in the 1880s and '90s. Nov 4, 1929. (F-29-159)

FIGURE 180.
We stopped at Mestia, the capital of Swanetia, both on the way to Ushgul and on the way back. There are many good examples of siege towers in Mestia as shown in this picture. During the interval between when the Bolsheviks first tried to get into Mestia in 1917 until they were successful in about 1924, the blood feuds began again. There was no central authority to control it. I met the Ispolkom Chairman, or local political boss, of Mestia who had spent four years in his family's tower for protection. During that time he was fed by the family from below using a ladder. Nov 9, 1929. (F-29-C213)

FIGURE 181.
Our host in Mestia was Said (pronounced Si-eed) Japaridze (R), a typical Caucasian— very proud. Even in these remote mountain districts, this man had that air of authority. I am afraid that later he was categorized as a kulak, a "rich" farmer in that he had an extra cow or extra room in his house. Many such "prosperous" people were persecuted (liquidated) by Stalin in the mid-30s. Muzahan, in the center, always dressed in Caucasian robes and carried a sword. Nov 8, 1929. (F-29-200)

FIGURE 182.
At the time of our visit in Mestia, this was the house where Muzahan had an apartment. (F-29-C205)

FIGURE 178

FIGURE 179

FIGURE 180

FIGURE 181

FIGURE 182

FIGURE 183

FIGURE 184

FIGURE 185

FIGURE 186

FIGURE 183.
The Church of the Virgin above Ushgul. This church had a twelfth century bible. To go into the church required that the three men in charge of the church be present when the door was unlocked and opened. I leafed through the bible and in the margins there were notations in ancient Georgian, from about A.D. 900. Only the priest could read it. I talked to him about translating it because there was some interesting local history there but communication was such that I never found out what it said. Nov 5, 1929. (F-29-C175)

FIGURE 184.
The churches in the area were really chapels. From what I could see, they were little treasure houses, keeping sacred ancient religious items. Not even the priest could enter its portals without permission of the village elders. These were not only relics of the area—places far away that were being invaded would send their treasures to these remote areas where they were out of the reach of the invaders. This church was near Mazaire. Oct 29, 1929. (F-29-C112)

FIGURE 185.
An old cross at the church in Laham near Mestia. Nov 8, 1929. (F-29-C199)

FIGURE 186.
An old icon of beaten silver also at the church in Laham near Mestia. Nov 8, 1929. (F-29-C198)

FIGURE 187

FIGURE 190

IN PRESENTING these few pictures, I realize that they fall far short of portraying the real beauty and grandeur of the region of the Central Caucasus, and that they cannot entirely convey the strange and mysterious qualities of the old Swanetian battlements. However, I hope these views will at least suggest the unique setting in which the gloomy history of this little group of mountaineers may be traced, and perhaps make it possible to attain a better understanding of their own problems as well as those which confront the governments who, at the present time, feel it their duty to help these people to raise themselves from barbarous ways of living and thinking to those of the twentieth century.

WM. OSGOOD FIELD

FIGURE 188

FIGURE 191

FIGURE 189

Foreign Relations

Alice was very interested in the treatment of women and children in the Soviet Union and joined me in Moscow for a brief visit after my first trip in '29, and then we traveled there together in '31. The Soviet Union appeared more socially enlightened than the modern world, with provisions for the education of women and women professionals. Many women in Russia were becoming doctors. Alice studied this treatment of women as well as children, and later she published a book entitled *Women and Children in the Soviet Union*.[11] Alice helped raise my social consciousness considerably. I was quite naive up to that point about politics and any theories of government.

As a result of going to the Soviet Union twice in '29, in '31 and another short trip in '33, as well as marrying Alice, I became concerned with the cultural relations between the United States and the Soviet Union on a deliberately nonpolitical basis. There was a crying need for a better understanding of the Russian people, rather than just the disagreement with and hostility towards their political system. It is the same problem we find today: we don't know enough about the way

of life of other people around the planet. It still is a battle in this world to achieve some understanding between people. So I worked over the years with the Society for Cultural Relations with the Soviet Union, which became the American-Russian Institute. It was founded in the mid-1920s by several people, many of whom had been in the diplomatic corps or in the art and theater world. Its focus was on cultural aspects. I never had the least interest in Communism. I also belonged to the Council on Foreign Relations, which is still a very interested group in New York City. The members were mostly from business, industry, and government. We were always suspect in later years, especially during Senator McCarthy's campaign against those with any connections with or interest in the Russians, but we avoided the political angle, and this worked quite successfully. It was a big group, several hundred people, some of them very interesting. I remember one evening at a banquet I was asked to join a table with about eight others, including Averill Harriman and Alan Dulles. As I recall, Dulles was head of the CIA at that time. Harriman remembered the trip to Alaska with his father in 1899, when he was about nine years old.[12]

FIGURE 192.
Alice and John in 1949.
(49-161; photo courtesy J.O. Field.)

FIGURE 193.
The whole family in 1943.
(No number; photo courtesy J.O. Field.)

FIGURE 192

FIGURE 193

FIGURE 194.
This picture was taken at Ladd Field late in the winter of 1943–1944 when it was a little warmer than when we arrived and when we had a few more hours of daylight. (No number; photo courtesy J.O. Field.)

FIGURE 195.
Planes at Ladd Field in late winter, probably March, 1944. A Lockheed Lodestar (top) and a C47, the military designation of the DC-3 (bottom). (No numbers; photos courtesy J.O. Field.)

FIGURE 196.
My crew at Ladd Field, 1944. (No number; photo courtesy J.O. Field.)

FIGURE 194

FIGURE 195

Fitzpatrick Travelogues

When I came back from the Caucasus, I went to work for James Fitzpatrick Travelogues from 1932 to 1935. Fitzpatrick had a contract with MGM for twelve one-reel travel films per year. In '33, I was sent to what was then British Guiana on the northeast coast of South America. It is now Guyana, an independent country. One of the main attractions in British Guiana was Kaieteur Falls, a high waterfall in the interior. It is a magnificent cataract in which the whole Potaro River pours over a cliff almost 750 feet high. It is one of the tallest, biggest-volume falls in the world (fig. 189). When I arrived in Georgetown, I had in mind the possibility of working out a trip to the falls, and the first thing the man from the local office of MGM who met me said was we wouldn't be allowed to leave until we had gone up to Kaieteur Falls. We went a long way by truck and then over a corduroy-type road through the jungle and then by boat up the river to as near as we could into the amphitheater at the base of the falls. The next day we went to the top and took pictures looking down, and it made a pretty good film sequence. So instead of spending two weeks in British Guiana, which I was supposed to do, I spent four weeks. After the film was made, I remember going to the Capitol Theater in New York. When the ten minutes of British Guiana came on, people clapped at the end of that. I'd never heard anything like this before. I sat through a second showing, to make sure this wasn't just my imagination. And I got the impression that the clapping was for the footage of Kaieteur Falls as the climax of our film.

FIGURE 196

Soon after, I found that I could only work as a director-cameraman outside the United States because the cameraman's union in the U.S. was pretty well closed to any new people unless they were relatives of current members. I didn't see any future in making short travel films taken abroad. By then I also had a family to consider. My daughter Diana was born in 1930 (to be followed by my son John born in 1940) (figs. 190–193). Thus ended my employment in the film industry, although my experience in photographing under difficult circumstances was utilized during World War II (figs. 194–196).[13] Fortunately during this period of working in the film industry, I was able to make two more trips to Alaska on my own to photograph and survey the glaciers, in 1931 and 1935, as my interest in that field continued to increase.

Notes

1. See Preface for an explanation of why the story of the development of Bill's career in glaciology is interrupted by this chapter.

2. Brown, "The Beach Anglo-Egyptian Sudan Expedition" (1929, 63–70).

3. See Freshfield, *The Exploration of the Caucasus* (1896); Mummery, *My Climbs in the Alps and Caucasus* (1895); von Dechy, *Kaukasus: Reisen and Forschungen im Kaukasischen Hochgebirge* (1905).

4. Douglas Freshfield (1845–1934), a distinguished British mountaineer and explorer, was perhaps the most interesting writer [about the Caucasus] in English. Merzbacher, a German, wrote about the same time as well as Moriz von Dechy, who wrote in Hungarian (William O. Field, personal communication, 1988).

5. See Allen, "New Political Boundaries in the Caucasus" (1927); Allen, *A History of the Georgian People from the Beginning down to the Russian Conquest in the Nineteenth Century* (1932); Allen and Muratoff, *Caucasian Battle Fields: A History of the Wars on the Torco-Caucasian Border, 1828–1921* (1953); Baddeley, *The Russian Conquest of the Caucasus* (1908); Baddeley, *The Rugged Flanks of the Caucasus* (1940).

6. Bill sent all his Caucasus and USSR material to the Golda Meir Library at the University of Wisconsin–Milwaukee to become part of the American Geographical Society Library Collection. The library was transferred there in 1980 when the Society moved to another location in New York City and decided to no longer maintain a library.

7. Field, W.O., "Travels and Mountaineering in the Caucasus" (1930).

8. In March 1917, the Bolshevik Revolution drove the czar of Russia from power, and by November the Bolsheviks (later called Communists), led by Lenin, seized the government (Dunn, *The Russian Revolution*, 1949).

9. Amkino was the organization that represented the Soviet film business in the United States and had the rights to show films about the Soviet Union. It distributed beautifully done and very dramatic films about the Bolshevik Revolution and the Soviet Union (William O. Field, personal communication, 1988).

10. Julien Hequembourg Bryan (1899–1974) was a famous photographer and lecturer who journeyed to all parts of the world to take pictures of people. He left Princeton in 1917 to serve at the front with the French ambulance service during World War I. The pictures he took and the lecturing he did for the government launched a career as a documentary photographer and popular lecturer. He first went to Russia to film in 1931. This was followed by several trips in the following decade (Block, *Current Biography: Who's News and Why*, 1940, 114–115).

11. Field, A.W., *Protection of Women and Children in Soviet Russia* (1932).

12. See Chapter 2, note 8 for a description of the Harriman Alaska Expedition of 1899 and the resulting publication.

13. See Appendix E for Bill's description of one of his assignments during World War II.

FIGURE 197

CHAPTER 6

BRINGING IT ALL TOGETHER
1931–1935

In any study of glacier fronts dated photographs are of prime importance, for they furnish accurate records and can be obtained when there is not time for detailed observations. If the photographs are taken from easily recognized stations which can be occupied in later years their value is still greater.

Grant and Higgins (1935, 10)

Prince William Sound

My next unofficial scientific trip to Alaska after my 1926 trip to Glacier Bay was in 1931. Will Wright[1] and H.F. Reid[2] planned to go back to Glacier Bay that year. They invited me to join them as I had been thinking about visiting Glacier Bay again. It had been five years since I had last been there, and I knew I would have learned a lot if I went to Glacier Bay with those two experienced men. But I also had been thinking of going to Prince William Sound.

Prince William Sound (map 27) had not been visited for glacier studies since Dora Keen's visit in

1914 and 1925 (map 28).[3] Before that there were visits only by Martin[4] in 1909, 1910 and 1911, Grant and Higgins[5] of the USGS in 1905, 1909 and 1910, and the Wright brothers[6] in 1904 and 1906. Fred Moffit[7] of the USGS was there in 1924 but I didn't know about that or see his pictures until I returned. Even with these visits by several people, there had been no systematic effort to map the glacier termini and determine whether they were advancing or receding.

I thought it more important that I visit a new place to investigate the behavior of the glaciers there, which was so different from that of the glaciers in Glacier Bay, than to go back to Glacier Bay. Therefore I decided to go to Prince William Sound[8] to start systematically photographing and surveying from Martin's stations, those that Grant and Higgins had established a few years earlier, and those established by Gilbert and Gannett of the Harriman Alaska Expedition in 1899.[9] By then, I knew a good deal more about locating old survey and photo stations and establishing new ones. Martin had coached me a lot on that. He always enjoyed telling me about these things so I could carry on. In fact, early that summer of '31 I spent a couple of days with Dr. and Mrs. Martin during their summer vacation on the coast of Maine, learning about surveying, what to look for, what to think about, and where his stations were.

FIGURE 197.
Columbia Glacier from the summit of Heather Island.
(F-31-14, 16)

FIGURE 198.
Here I am at Columbia Glacier Sta. 9, Sep 28, 1931. (F-31-163)

FIGURE 199.
The west margin of Columbia Glacier from Sta. G. Top photo by G.K. Gilbert (No. 355) on the Harriman Expedition in 1899; center photo by U.S. Grant (No. 91) of the National Geographic Society Expedition in 1909 showing an advance in progress; bottom photo I took in 1931 showing the barren zone uncovered by the retreat which followed the advance. (First two photos from copy photos in the WOF Collection; last photo F-31-59.)

FIGURE 200.
View WNW from the west margin of Columbia Glacier to the cabin left from mining days that we stayed in. (See maps 29 and 30 for location.) (F-35-320)

FIGURE 198

FIGURE 199

Columbia Glacier (fig. 197) was the first glacier we visited and we spent what now seems like an inordinate amount of time for what we accomplished there. There are extensive tide flats there and sometimes we lost days when we couldn't launch the boat because the water was too shallow and the mud too deep to move it from its mooring. We made a survey (fig. 198) of the entire terminus, which was a wide one, about five miles from one side to the other. Such a survey had not been done since Martin's day in 1910 (map 29). We found that Columbia Glacier recently had begun to retreat and was only 300 feet from where it had advanced into the forest and formed a trim line in old trees (map 30; fig. 199). Later, Cooper[11] dated some of those trees, and found that they were over 400 years old. This was sort of unheard of at that time—to find glaciers that had reached maximums in this century when the Little Ice Age was supposed to have ended in the middle of the nineteenth century.[12] So our 1931 trip established the greatest advance of Columbia Glacier, which we estimated to have occurred around 1920.[13]

Mining records from Prince William Sound show that activity around Columbia Glacier began around 1910. There were mines up both sides of the glacier and relics from those mines were still there in 1931. We stayed in a hut miners had built on the west side of Columbia Bay (fig. 200). It was very close to where the glacier's maximum position had been in 1920, so I think it predated that advance. Also on the west side there was a tote road that went up through the woods, to get past the terminus and out onto the ice, where it flattens out. The mine was called Billy Hole Lake. Even by '31 you couldn't use the road because the ice had lowered so much that there was a cliff where they used to step out onto the glacier. This same situation happened at Valdez Glacier. There had been mines along both sides of the glacier (fig. 201), but the surface of the glacier had lowered so much that it became difficult to reach the buildings.

FIGURE 200

FIGURE 201   FIGURE 202

Besides discovering that Columbia Glacier had recently been advancing, we found Harvard Glacier in College Fiord still advancing into new territory (figs. 202, 203). There weren't many trees in front of Harvard Glacier, but there were one or two that subsequently turned out to be several hundred years old. We also found Harriman Glacier advancing since the last reports of it (map 31; fig. 204), and Meares Glacier advancing into forest. In a few cases, there was recession. So we had a totally new picture[14] of something to be followed up. People speak of Glacier Bay as being my favorite place, but it certainly has to share with Prince William Sound, because the behavior of the glaciers in the two areas is so different.

Unknown to me until later, the [U.S.] Geological Survey had sent C.K. Wentworth and Louis Ray,[15] both geologists, to Alaska that year. They had made a survey of glaciers along the coast, but our surveys overlapped only at Valdez Glacier. Their report was published by the Geological Society of America[16]—excellent report. Ray liked that country up there, and I remember some years later how dismayed he was because the Survey was sending him down south to study the Mississippi Delta.

H.F. Reid and C.W. Wright went into Glacier Bay in 1931 as planned, and took a lot of pictures and made a few rough plane-table maps (fig. 205). They reoccupied our Muir Inlet Sta. A that we'd set up in 1926. This became Wright's Sta. 3 and in subsequent years our Sta. 3. They also went into Johns Hopkins Inlet, which we could not go into in 1926 because of the ice. So they were the first to go into the inlet specifically to study glaciers, and they made a map of it (map 32). Wright wrote up a report of what happened between his visit in 1906 and 1931 but it was never published. It's on open file status with the USGS, but Wright gave me a copy of the manuscript with his pictures and maps.[17] Cooper had been in Johns Hopkins Inlet in 1929, and, as far as I know, he was the first one to actually take a boat into that inlet and take some

pictures, although he didn't map. In addition, the navy took some air photography in 1929, which was excellent—superb work.[18] However, I was glad I'd been to Prince William Sound and I've never regretted my decision to go there instead of Glacier Bay.

"Glaciers of Northern Prince William Sound"

Besides the 1931 trip being my first visit to Prince William Sound, it also was significant because when I came back, Martin wanted me to give an illustrated talk about my trip at the Association of American Geographers' annual meeting in late December in Ypsilanti, Michigan. I prepared lantern slides, which in that day were the three and a half by four inch glass slides. Although the photography was all black and white, there were places where you could have the slides colored. I sent mine to a firm in Jersey City where some older ladies painstakingly put color on these slides. They sometimes tended to paint a sunset in places where there shouldn't be one, so you had to tell them not to put a sunset here because that was facing east or north. I still have those colored slides, which may be among the most stable of photographic records. About the twenty-ninth of December 1931, I went out to Ypsilanti and I was scheduled to talk first one morning. There probably had been a banquet the night before. I think there were only four people in the audience when I got there. I wasn't dismayed. The man who was chairman of the meeting was Isaiah Bowman.[19] I'd never met him before but I'd heard of him, as he had been the director of the American Geographical Society. He introduced me and I gave the talk. It went reasonably well. A few more people had wandered in by the time I'd finished. A few weeks later, I received a letter from Dr. Gladys Wrigley,[20] the editor of the American Geographical Society publication *Geographical Review*, who asked me if

FIGURE 201.
The Ramsey Rutherford Mine, Valdez Glacier, probably early 1900s. [Photo possibly taken by B.L. Johnson (USGS) between 1913 and 1917. Ed.] (Photo in the WOF Collection.)

FIGURE 202.
The east margin of Harvard Glacier terminus looking ENE from the boat. The creek is from Downer Glacier. In 1914 the ice front was about 1300 feet away from the creek. (F-31-372)

FIGURE 203.
*The west margin of the
Harvard Glacier terminus
viewed from the boat. We
found the ice advancing
into mature alders. Oct 7,
1931. (F-31-362)*

FIGURE 204.
*We found that Harriman
Glacier was also advancing.
View SW to the fresh
moraine at the south end of
the terminus. (F-31-518)*

FIGURE 205.
*Charles Will Wright (R)
and Harry Fielding Reid
(L) in front of float plane
prior to leaving on a
flight, August 5, 1931.
(Photographer unknown;
copy photo from the WOF
collection.)*

FIGURE 203

FIGURE 204

FIGURE 205

I would consider writing up my talk. I learned later that the editors very seldom asked anybody to write a report, because then they were stuck with it, whether they like it or not. Dr. Wrigley was very meticulous and went through everything with great care. She somehow survived my efforts and produced my article "Glaciers of Northern Prince William Sound."[21] It was the closest thing I had come to a technical professional report. It was well received. This is how I began to have contact with the American Geographical Society. I was in and out of the society quite a lot while Dr. Wrigley helped me edit my paper, and this was really the entering wedge for eventually joining the staff of the society. A year later I was asked by the American Alpine Club to write an article on Prince William Sound, which, up till then, very few people had written about. The article was entitled "Mountains and Glaciers of Prince William Sound."[22] Writing these papers and having them published made it easier for me to deal with the geographers and geologists in Washington who had visited Alaska before, because they saw I was serious about studying the glaciers. Martin especially was pleased with me.

Another significant result of the 1931 trip was for me to meet Dr. François E. Matthes, senior geologist at the USGS.[23] At that time he was involved in the formation of a group in the U.S. concerned with the study and reporting of glacier variations in key areas in the U.S. as part of an international effort to coordinate the information obtained in the past, and to promote a program of continuing systematic observations.[24] I began working with that group, the Committee on Glaciers, in 1931 and became chairman of the committee in 1948.

Establishment of a Glacier Monitoring Program

My 1926 trip to Glacier Bay and 1931 trip to Prince William Sound, followed by the two publications, set the stage for looking at glaciers in both the Glacier Bay area of the Coast Mountains and Prince William Sound. After the 1931 trip, I met William S. Cooper,[25] head of the department of botany at the University of Minnesota. This was my first association with a scientist other than a geologist whose research was related to glaciers. He'd first been to Glacier Bay in 1916, and again in 1921 and 1929. His interest was in knowing the position of the ice edge at different times so he could study the growth of vegetation and plant successions in the newly exposed ground as the glaciers retreated. From this he could learn how long it took the vegetation to begin to grow after the ground became ice-free. There is always a period of adjustment as the ground often oozes moisture for several years and nothing can grow, then suddenly it dries up and plant life begins, very rapidly in the case of Glacier Bay. This study of vegetative succession continues to date. Cooper's successor, Donald B. Lawrence, still works in Glacier Bay occasionally, and Mark Noble of the University of Minnesota and Ian Worley of the University of Vermont carry on most of the work. So the mapping of glacier termini fit into Cooper's studies and thus we made plans to go to Alaska together in 1935. Cooper had not been to Prince William Sound at that point, as all his work had been in Glacier Bay, so this trip we decided to do both. It was my first return to Glacier Bay, nine years after I'd first been there, and it was Cooper's fourth trip.

We had two assistants. Brad Washburn, a well-known Alaska mountaineer who later took very high-quality aerial photographs of the mountains in Alaska, recommended two Dartmouth College students, who had been with him the year before on his expedition to Mount Crillon. One was Russell Dow. He had just finished his freshman year at Dartmouth. His father was an engineer on the Maine Central Railroad, which provided Russ with a pass to go anywhere in the United States by rail for free. So he got to the Pacific Coast with no expenses except food, and that was a big help to me. It turned out he liked Alaska so much, and didn't like college and academic work, that he just moved up there and now lives in Palmer. The other was Bob Stix, a New Yorker who subsequently went into business in New York. I've seen him only a couple of times since. On that trip we had Tom Smith of the motor vessel *Yakobi* as skipper. He had taken Cooper up on two or three of his previous expeditions (figs. 206–209).

Cooper worked in lower Glacier Bay the week before I joined them, making the first really detailed studies of the vegetation on the terminal moraines. Everyone knew the moraines at the lower end of Glacier Bay, which had been formed by the maximum advance of the Glacier Bay glaciers, were there but nobody knew their age. Cooper determined fairly definitely that the glaciers reached their maximum in the middle of the eighteenth century and formed the moraine and the outwash plains one sees now (map 33).[26] My one basic objective of the trip was to map Muir Inlet[27] (map 34). It had changed a great deal since the last survey had been made in 1907 and published on the boundary survey map issued in 1928.[28] I also wanted to add to the data from our survey of '26. I was not able to complete the survey in Muir Inlet as the rain never stopped and the fog never lifted. But we did better in the other inlets. We surveyed in Johns Hopkins Inlet and

FIGURE 206

FIGURE 207

FIGURE 208

FIGURE 206.
Russ Dow in the skiff in Surprise Inlet. The view is toward Toboggan Glacier. (F-35-600)

FIGURE 207.
The Yakobi *in Fords Terror.* (F-35-242)

FIGURE 208.
Tom Smith in the pilothouse of the Yakobi. (F-35-250)

FIGURE 209.
Tom Smith (L), Russ Dow, and Ed, the cook (R) on the Yakobi. *(F-35-248)*

FIGURE 210.
Cooper on Sta. 1 in upper Johns Hopkins Inlet, with Gilman Glacier in the distance. [Johns Hopkins Glacier advanced over this station about 1976. Ed.] (F-35-73)

FIGURE 211.
The M/V Fidelity *in Icy Bay, Prince William Sound. (F-35-841)*

FIGURE 209

FIGURE 210

FIGURE 211

continued the work that Wright had started in '31 (fig. 210). In the survey, we located the small hanging glaciers that Sargent, chief topographer for the Geological Survey, had located from the 1929 aerial photography by the U.S. Navy,[29] and we subsequently gave them names.[30] In Tarr Inlet we had much the same bad weather as in Muir Inlet but we were able to survey in Hugh Miller, Geikie, and several other inlets (maps 35, 36).[31]

At the beginning of the trip, I called Cooper "Doctor" Cooper but he said to call him Bill. And I remember Ed the cook saying how disgraceful it was for us young fellows to call him Bill. Cooper was a man of great breadth of intelligence, besides being a good botanist. He was a great lover of music and an admirer of his friend, the director of the Minneapolis Symphony Orchestra. I also found out we had a common bond during one of his visits with me in New York. On the walk back to the office from lunch, a lot of fire trucks went by and I knew that some of them came from a long distance away, which meant that there was something really unusual going on. I was thinking how I'd like to go down to see the fire, but guessed I better get back to my desk. I mentioned returning to the office to Cooper and he said: "I don't care what you're going to do, I'm going to the fire." So he was just like me. He would always follow a fire engine. We often referred to the subject in subsequent meetings.

After Glacier Bay, the four of us went on up to Prince William Sound with a brief stop at Cordova where we went up to Miles, Childs and Allen glaciers.[32] In Valdez we picked up a boat we had chartered, the M/V *Fidelity* (figs. 211–213). In those days the boats you chartered were usually fishing boats and you had to wait till the end of the fishing season, around the fifteenth of August, before you could hire one. We visited Valdez, Shoup, and Columbia glaciers, continued west to Meares Glacier, and then on to College Fiord and Harriman Fiord. Cooper then returned home for the opening of college in the middle of September. Dow, Stix, and I extended the work to the western side of Prince William Sound in the Kenai Mountains, into Passage Canal, Blackstone Bay, Kings Bay at the head of Port Nellie Juan, and Icy Bay, thus covering the part of Prince William Sound that I hadn't visited before (map 27). And so in '35 we were probably the first to go into these areas and study some of those glaciers since the USGS party of Grant and Higgins from 1905 to 1909.[33] This work was also the beginning of widespread repeated measurements of glaciers in the Prince William Sound and Kenai Mountains area. When we returned to Cordova, we took a flight around Prince William Sound. The pilot's name was Kirkpatrick. He later was killed on a rescue mission in the area. He had a Bellanca as I recall (fig. 214). At that time, as now, they took off and landed on Eyak Lake just behind Cordova. This was the first time I'd taken any aerial pictures (fig. 215). They are not of very good quality, but they show the features we were interested in.

Now in 1935 the Matanuska Colony was being organized.[34] President Roosevelt had conceived the idea of sending some of the families who were on relief in Cleveland and Detroit and farmers in Minnesota and Wisconsin to Alaska to farm in the Matanuska Valley near the town of Palmer, northeast of Anchorage. They received some subsidy for this massive move to an entirely new environment. I was interested in seeing the colony as part of the modern story of Alaska. In Cordova, after we had finished our work, we accidentally met the minister of the local church in Matanuska. He was not going to be there for another week, and offered us the use of his tent. Bob Stix had returned home, so Russ Dow and I headed to Palmer and the Matanuska colony. We spent about three days there and took black and white films and still photos of the colonists building houses and developing their new surroundings, so there was quite a bit of activity going on.[35]

We continued on to Juneau where we had about a week to spend. We hired Tom Smith again in the *Yakobi* to take us into the fiords of southeastern Alaska (map 10). Other than Taku Inlet, this was my first visit to those fiords. Tracy and Endicott arms are the main ones. They are beautiful and the glaciers at the head of those fiords are as interesting as those in any other fiords along the coast of Alaska. I knew about this area because Cooper had been there and I had reviewed the literature on them. We reoccupied some of the stations that had been occupied by Buddington of the Geological Survey, and set up new ones.[36] At the end of this second trip out of Juneau, we included Tom Smith on a flight over Glacier Bay. I think that it was his first airplane ride. During the course of the two trips in the boat with us that summer, Tom had mentioned that he probably knew every rock in the bay. From the plane, he looked down and saw all the reefs he'd been trying to avoid, and some he hadn't avoided. He had quite a trip. That was the first time I'd taken pictures of Glacier Bay from the air (fig. 216) and they were, I think, the second series of aerial pictures of glaciers in Glacier Bay, following the superb results of the navy's flight in 1929. Those two flights in the summer of '35 were my first aerial experience, and strangely enough, I never had another flight over Glacier Bay again until 1982.

FIGURE 212

FIGURE 213

FIGURE 214

FIGURE 212.
Frank Granite (L), the engineer, and Les Newton (R), the cook, on the deck of the M/V Fidelity. *(F-35-853)*

FIGURE 213.
William S. Cooper on the deck of the M/V Fidelity. *(F-35-551)*

FIGURE 214.
Kirkpatrick's Bellanca seaplane in Eyak Lake near Cordova that we used for the flight over Prince William Sound. That was my first photo flight. (F-35-855)

FIGURE 215.
*Aerial view of Columbia
Glacier in 1935 taken while
on my first photo flight.
The glacier terminated on
Heather Island at that time.
(F-35-876)*

FIGURE 216.
*Aerial view of Muir Glacier
terminus I took in 1935 on
my first flight over Glacier
Bay. The arrow on the left
indicates the location of
McBride Glacier where it
flows into the Muir, and the
arrow on the right points to
the Nunatak. (F-35-121)*

The '35 trip ended with my writing "Observations on Alaskan Coastal Glaciers in 1935," with some maps and the changes we noticed that had occurred over the four years since our previous visit.[37] That was my second publication in the *Geographical Review.* Cooper wrote his piece, "The Problem of Glacier Bay, Alaska: A Study of Glacier Variations,"[38] which is a classic. He cited the vegetation that indicated where glaciers terminated at an earlier time. He also discussed the history of the establishment of Glacier Bay National Monument, with which he had been deeply involved as a member of the Ecology Society of America. I think he was the one who actually proposed setting the area aside as a national monument after his trips there in 1916 and 1921.[39]

So, even though I was not a professional in the field of geology or glacier studies, I was meeting and working with some of those who were. I also was becoming acquainted with the folks at the American Geographical Society, where I began my professional career in glaciology a few years later.

FIGURE 215

FIGURE 216

Notes

1. For a biographical sketch of Will Wright, see Chapter 3, note 16.

2. For a biographical sketch of Harry Fielding Reid, see Chapter 3, note 33.

3. Dora Keen (Handy) (1871–1963) was an extensive world traveler. She made many difficult ascents in the Alps, and the first ascent of Mount Blackburn (16,140 feet) on May 19, 1912, during which she made important observations of snow fall. In 1914 she made the first exploration of Harvard Glacier (in Prince William Sound) to its source. She is the author of several magazine articles cited in the bibliography.

4. For a biographical sketch of Lawrence Martin, see Chapter 3, note 22.

5. During the summer of 1905, Ulysses Sherman Grant (1867–1932) was engaged in a study of the ore deposits and the general geology of Prince William Sound for USGS. Grant and D.F. Higgins continued the work in 1908. In 1909 Grant and Higgins extended the work to the southern part of the Kenai Peninsula. They made notes and maps and took photographs of all the tidewater glaciers and many of the other glaciers near tidewater. This was secondary to the study of the bedrock geology and ore deposits, however Grant and Higgins thought it "worthwhile to put on record the information thus obtained regarding the glaciers, for it will afford a basis for future study of the fluctuations of these ice streams" (Grant and Higgins, *Coastal Glaciers of Prince William Sound and Kenai Peninsula, Alaska,* 1913).

6. For a biographical sketch of the brothers C.W. Wright and F.E. Wright, see Chapter 3, note 16 (The editor can find no evidence to confirm the Wrights' visit to Prince William Sound in 1904).

7. See Chapter 3, note 36 for a biographical sketch of Fred Howard Moffitt.

8. See Appendix C for a summary of the trip.

9. Grove Karl Gilbert, geologist (see Chapter 2, note 12), and Henry Gannett, chief geographer, USGS, were members of the Harriman Alaska Expedition. See Chapter 2, note 8 for a description of the expedition and the resulting publication.

10. Tarr and Martin, *Alaskan Glacier Studies* (1914).

11. Cooper, "Vegetation of the Prince William Sound Region, Alaska; with a brief excursion into post-Pleistocene climatic history" (1942).

12. Little Ice Age refers to the maximum worldwide ice extent in the middle of the eighteenth century. During the Little Ice Age, approximately A.D. 1450–1850, mountain glaciers all over the world advanced considerably beyond their present limits (Imbrie and Imbrie, *Ice Ages*, 1979, 181.)

13. Field, W.O., "The Glaciers of the Northern Part of Prince William Sound, Alaska" (1932, especially 375).

14. See note 13 above, pages 361–388, for Bill's complete report of his trip to Prince William Sound.

15. For a biographical sketch of C.K. Wentworth and Louis Ray, see Chapter 3, note 17.

16. Wentworth and Ray, "Studies of Certain Alaskan Glaciers in 1931" (1936).

17. See Appendix E for Bill's description of the report and the impact on Bill of Wright's unlabeled photographs from his trips.

18. See chapter 8 for a description of the navy's air photography from 1929 included in an overview of the history of glacier photography in Alaska.

19. Isaiah Bowman (1878–1950) was president emeritus of Johns Hopkins University and one of the world's leading geographers. Bowman became the foremost authority on political geography in the country and is described as perhaps the most efficient coordinator of topography, population pressures and economic resources the world has ever known. Bowman was a member of the American delegation to negotiate the Treaty of Versailles, and participated in post–World War II conferences that led to the creation of the United Nations. Bowman was the first director of the American Geographical Society (1915–1935) and president of Johns Hopkins University (1934–1948) (*New York Times*, Jan 7, 1949).

20. Gladys M. Wrigley received her doctorate under Isaiah Bowman at Yale University and became the first woman to receive a Ph.D. in geography from an American university. She served as editor of the *Geographical Review* for three decades and became one of the most influential scientific editors of the first half of the twentieth century. The journal was to present research of high intellectual quality in a style that was broadly accessible for a readership of geographers and nongeographers. The body of principles and strategies that she used to achieve these goals continues to be a source of guidelines for any scholarly journal that seeks to communicate beyond a small, specialized readership. (McManis, "The Editorial Legacy of Gladys M. Wrigley," 1990.)

21. Field, W.O., "The Glaciers of the Northern Part of Prince William Sound, Alaska" (1932).

22. Field, W.O., "The Mountains and Glaciers of Prince William Sound, Alaska" (1932).

23. François Emil Matthes (1874–1948). Born with twin brother Gerard Hendrik Matthes in Amsterdam, the Netherlands, on March 16. The boys were schooled in Switzerland and later Germany. Their father's hobby was mountaineering and he taught them to use and interpret topographical maps. The twins also received instruction in drawing and François revealed a great talent for art. The twins emigrated to America in 1891, graduated from Massachusetts Institute of Technology in 1895 with honors, and became U.S. citizens the next year. In 1896, François joined the USGS. From 1930 until his retirement in 1947, he was senior geologist. For many years, he expressed his talent for drawing, as well as for precise, analytical, scientific work, in the making of topographical maps. He became a master of reconnaissance mapping, first topographical and later geological, much of it carried out in exceedingly difficult terrain without benefit of roads, or sometimes even trails. In June, 1905 Matthes began mapping Yosemite Valley for the survey. His last major field assignment came in 1910 when he was put in charge of mapping Mount Rainier National Park. His first geological report was *The Geologic History of the Yosemite Valley*, Professional Paper (1930). It was immediately hailed a classic by Kirk Bryan. Many of his later years were spent involved in scientific organizations. He was chairman of the Committee of Glaciers of the American Geophysical Union (AGU), which included

preparing yearly reports summarizing and analyzing data collected on glaciers in the U.S. These were published annually from 1932 to 1946 in the *Transactions of the AGU*. He was a member of the International Association of Scientific Hydrology, as well as secretary of the International Commission on Snow and Glaciers, a division of the association. He retired in June 1947 and died in June 1948. (Fritiof, "Memorial to François Emile Matthes," 1956.)

24. This study emerged from the formation of the International Glacier Commission, organized at the International Congress of Geologists at Zurich in 1894. Its objective was to make observations of the changes occurring in the length and thickness of glaciers. Much information had already been collected on the variations of the glaciers in the Alps, so "it was deemed desirable to know something of the variations of glaciers in other parts of the world, to determine whether these variations are synchronous on different continents and on opposite sides of the equator. To what extent the variations of glaciers are dependent on meteorological changes, and to what extent on the size and shape of reservoirs, etc., is recognized as a problem whose solution is hoped for" (Reid, "Variations of Glaciers I," 1896). The two representatives from the U.S. were H.F. Reid of Johns Hopkins University and William H. Hobbs, professor of geology at the University of Michigan. The activities of the International Glacier Commission came to an end in 1914 when World War I broke out. It was not until 1927 that a new international commission was formed as part of the International Union of Geodesy and Geophysics. Through Matthes' efforts a Committee on Glaciers in the U.S. was formed in 1931 in the Section of Hydrology of the American Geophysical Union. (William O. Field, personal communication, 1988.)

25. For a biographical sketch of William Skinner Cooper, see chapter 3, note 10.

26. Cooper, "The Recent Ecological History of Glacier Bay, Alaska" (1923, 100); Cooper, "The Problem of Glacier Bay, Alaska: A Study of Glacier Variations" (1937, 47).

27. Cooper, "The Problem of Glacier Bay, Alaska: A Study of Glacier Variations" (1937, especially figs. 10, 17, and 19).

28. For a description of the boundary survey work, see Appendix E.

29. See Appendix E for Bill's story of tracing some ice fronts from Sargent's maps.

30. See Appendix E for the text describing the names.

31. See Appendix C for a summary of the trip.

32. See Appendix C for a summary of the trip.

33. The description of the Grant and Higgins trip is given in note 5 above.

34. In April of 1935, 400 single men and 200 families from northern Michigan, Minnesota and Wisconsin were provided transportation to the fertile Matanuska Valley north of Anchorage, Alaska, by the Federal Government and each given a forty-acre tract. The government initially provided food; tools for clearing the land; for planting and for building homes; and livestock—cattle, sheep, hogs, and poultry. (*New York Post*, April 8, 1935.) See Miller, O.W., *The Frontier in Alaska and the Matanuska Colony* (1973).

35. The photographs are stored in the collection with all the other photos from Bill's 1935 trip to Alaska.

36. For a biographical sketch of Arthur F. Buddington, see chapter 3, note 9.

37. Field, W.O., "Observations on Alaskan Coastal Glaciers in 1935" (1937).

38. Cooper, "The Problem of Glacier Bay, Alaska: A Study of Glacier Variations" (1937).

39. Cooper, "A Contribution to the History of the Glacier Bay National Monument" (1956).

A GLACIER
STUDY PROGRAM
1940–1956

The American
Geographical Society

The American Geographical Society (AGS) had a long history of interest in the polar regions.[1] It sponsored various expeditions and published their accounts beginning when the organization was first formed in 1851. A number of the people who formed the Society were concerned with the search for Sir John Franklin, who had disappeared in the Arctic in the 1840s.[2] Of course, during the searches, people were exploring new areas, so from the very beginning the Society was interested in polar exploration, and the Society's library was rich in polar and natural science literature.

I joined the staff on December 2, 1940. My desk was in an alcove in the exhibition room because all the offices were taken. I was also given a bookcase and a chair, and a wastepaper

basket, I guess. Part of my introduction at AGS was Walter Wood, who had joined the Society a year or two before. I'd met Walter at an Explorers Club meeting in the '30s and we became good friends (fig. 217). He founded the Department of Exploration and Field Research at the AGS. He had climbed in the St. Elias Range and was interested in the development of expeditionary techniques.[3] His particular interest was air supply, because normally in these remote areas an expedition by dog team was required a few months to a year before a major expedition, to establish supply caches up the glaciers to be traveled by the main expedition.

Presently I went to see the director, Jack Wright. He later wrote the history of the AGS.[4] I asked him what I should do and he replied, "Well, you have published on the subject of glaciers, and I think it would make sense for you to develop a glacier-study program." That was the last

FIGURE 217.
Visiting with Walter Wood on Steele Glacier in 1967. Walter always stood very straight—military style— because of his permanent stiff neck. (F-67-K461)

FIGURE 218

FIGURE 219

FIGURE 218.
"Donald B. Lawrence at one of his quadrants near Johns Hopkins Station 3 for the study of the progression of plant life after deglaciation, Aug 13, 1950. In 1907 this area lay beneath 1750 feet [534 m] of ice and has been ice-free since sometime before 1926. The study of the invasion of plant life began in 1929 by W.S. Cooper." (From a poster display; F-50-K367)

FIGURE 219.
Paul Livingston and I at Hopkins Sta. 4. View to the northwest toward Topeka Glacier. Aug 15, 1950. (F-50-R448)

directive I ever received. I proceeded on that for the next forty years.

In the beginning there was no money to do much, except operate my desk. I worked alone with a part-time secretary until 1948 when I hired Marie Hatcher (Morrison), who was my full-time secretary for many years. The first project I started was a gazetteer of Alaska glaciers. This was to include the geographical setting and the history of the observations at each glacier, what was known of its variations, and references to maps and the related literature. I had the AGS library at my disposal, which included the invaluable map collection and the accounts by people, mostly from the Geological Survey, who had been to Alaska. I knew a few of them. I started with the glaciers in southeast Alaska and worked northwestward up the coast. I got as far as Taku Inlet with the manuscripts when World War II came along and I was away from September '42 until January '46. I picked up the gazetteer project when I came back, but soon it was crowded out by other activities.

During the war, Walter Wood was the military attaché in Ottawa for the Canadian government. He continued in that capacity after the war, and so the AGS asked me to take over as director of the Department of Exploration and Field Research. From then until my formal retirement in the mid-60s I managed this department.

Completing Unfinished Business

In 1941 I saw an opportunity to return to Alaska and finally map the lower end of Muir Inlet, something I had wanted to do since 1926.[5] All the maps were obsolete, as the glacier had retreated several kilometers,[6] the surface had lowered hundreds of meters, and there were a lot of new nunataks and ridges that previously had been under ice and thus were unmapped. The remapping of Muir Inlet (map 37) was the most important aspect of that trip and the map, as well as a description of the changes in the glaciers from 1880 to 1946, was published in 1947.[7] That report was well received and a bit of a breakthrough for me because my work was now being taken seriously.

Cooper couldn't come with me in 1941 to repeat his work of 1935, so he sent his assistant Don Lawrence (fig. 218). Don joined me again in Glacier Bay in 1950, as well as Paul Livingston (fig. 219). Don and I had two assistants in '41. Maynard Miller, who had just finished his freshman or sophomore year at Harvard, was my assistant. He had been with Brad Washburn on Mount Bertha in the Fairweather Range the year before, and Brad had recommended him as a good, enthusiastic worker. Tony Ladd helped Don Lawrence. Tony was the son of Dr. Ladd, who had organized the Mount Fairweather climb in 1931.[8]

Before the '41 trip I met Frank Heintzleman, the regional forester for the U.S. Forest Service office in Juneau.[9] He was a very nice man who thought we were doing something useful, and he was interested in helping with our study of glaciers. He arranged for us the use of the U.S. Forest Service boats the *Forester* in '41 and the *Ranger 10* in '50. I remember Captain Akins of the *Forester* couldn't understand what the hell we were doing. He liked to stay away from the ice and sandbars, and here we were going ashore at the glaciers' fronts to take pictures. In '50 the boat did go high and dry once. It was one of those cases where the boat swings into shallow water while at anchor. The tide came in and floated us off, so it was not a dangerous situation. We took pictures and laughed about it, all except the skipper (fig. 220).

Juneau Ice Field Research Project

Back at the AGS in 1946 Maynard Miller and I, while looking at pictures of glaciers, recalled mention by someone that the answer to what is the state of health of a glacier is found in its upper source regions, not at the terminus. The activity at the terminus is just the result. That led to the development of the Juneau Ice Field Research Project (JIRP), which began observations on Taku Glacier in 1949 following a reconnaissance expedition in 1948 (maps 10, 38). I did the deskwork for JIRP, and Miller and I collaborated on the fieldwork. From 1950 on Cal Heusser was head of the project, with Art Gilkey serving as field leader, followed by Ed LaChapelle in 1954, '55, and '56, and then Larry Nielson in later years. The work of the Juneau Ice Field Research Project on Lemon Creek Glacier north of Juneau was begun in 1953, and studies on Taku Glacier were suspended after the summer of 1953 because of that glacier's huge extent and complex nature. The project was conducted under contract with the Office of Naval Research, with logistical support from the U.S. Army and Air Force (figs. 221–238).

A lot of young people who were trained in those early years found out whether they were interested in continuing in this type of work after they had been through a season on the icefield, tramping over the snow and making measurements in the cold. One of them is Mel Marcus, who is now at Arizona State University and is the vice president of the American Geographical Society.[13] Many of the people who went to the Antarctic during the International Geophysical Year (IGY) in '57 and '58 had been trained by JIRP. So although JIRP served a purpose, it was a strain on our small office staff to keep going. Every year we had to make an application for funds, and follow it up with results. Miller gradually took over JIRP[14] under the auspices of Michigan State University around the time of the IGY.

In assessing the scientific contributions of JIRP, I think everybody has a slightly different opinion. At the AGS, we started out taking measurements and we published many reports.[15] Then the student group became too big of a group to train. You can train young people and teach them what the techniques are, but it spreads the scientific results much more thinly. And I think that now it's more of a training ground than it is scientific research. Miller still writes up the results of his JIRP studies, and although I don't always agree with the results, his enthusiasm is one of the things that has made it possible for him to keep going.

FIGURE 220

FIGURE 221

FIGURE 222

FIGURE 223

FIGURE 220.
The U.S.F.S. Ranger 10 aground on the tide flats in Holkham Bay in 1950. (F-50-R61)

FIGURE 221.
"Field headquarters, Camp 10, of the Juneau Ice Field Research Project. The research station is built on a nunatak [rock island] at an elevation of 3875 feet [1182 m] overlooking Taku Glacier, and located sixteen miles [25 km] upglacier from the terminus. It was built in 1949 and served as the main base for various satellite camps dotted over the glacier and tributaries. Aug 31, 1950."[10] (F-50-R627)

FIGURE 222.
Some of the Juneau Icefield personnel in 1952. Left to right: Robert Schuster, glaciologist; Ed LaChapelle, glaciologist; George Argus, surveyor; Art Gilkey, leader. (No number)

FIGURE 223.
At Camp 10, Aug 30, 1950. Left to right: Art Gilkey, died on K2 in 1953; Mel Marcus, now professor of geography and environmental studies at Arizona State University and on the Board of Directors of the AGS; Cal Anderson, drill rig operator from the Longyear Co; Bob Nichols (rear), Tufts University; Jerry Wasserburg, came with Henri Bader and later joined the Geophysics Department at Cal Tech; Bob Forbes, later Chair of Geology Department at the University of Alaska and State Geologist; Fred Milan, who became professor of anthropology at Wisconsin and then at the University of Alaska Fairbanks, specializing in Eskimos; Charles "Bucky" Wilson, later geophysics professor at the Geophysical Institute, University of Alaska Fairbanks; Cal Heusser, American Geographical Society and NYU Department of Biology; Paul Livingston, mountaineer in Portland, Oregon; and myself. (F-50-R583)

FIGURE 224.
Fred Milan at work, 1950.
He was known as Dr.
Mukluk because he studied
penguins in the Antarctic.
(F-50-R612)

FIGURE 225.
Maynard Miller,
glaciologist, making
observations of snow crystals
on the Juneau Icefield in
1950. (F-50-R672)

FIGURE 226.
"After a night in the snow,
surveyor George Argus
eats his breakfast before
"weaseling" off to another
glaciological check-point on
the icefield. The expedition
uses these weasels to speed
up the gathering of scientific
data over the 700 square-
mile [1813 km²] area of
the Juneau Icefield. Stored
on the glacier during the
winter, this weasel had to be
dug out from 30 feet [10 m]
of snow when the expedition
returned early this summer."
(From a poster display;
F-50-no number.)

FIGURE 227.
The Stanford Research
Institute (SRI) seismograph,
July 1949. Thomas Poulter[11]
is on the right. (B-160)

FIGURE 228.
George Argus surveying on
the icefield. (Photo courtesy
Ed LaChapelle, no. 780,
1952.)

FIGURE 224

FIGURE 227

FIGURE 225

FIGURE 228

FIGURE 226

FIGURE 229

FIGURE 230

FIGURE 231

FIGURE 229.
I came to know Dick Hubley[12] very well during the IGY. He was coordinator for the U.S. IGY glaciological program for the Northern Hemisphere. I admired him for what he knew, and how he was going about the study of the relationship between climate and fluctuations of glaciers. Dick was young and well educated for this sort of work: he knew his geology; he knew his climatology; he knew about the physics of the atmosphere. Micrometeorology was his specialty and he was described as this country's foremost authority in this field. Many of us thought that he would probably be the first person to become a professor of glaciology in this country. We were shocked to hear in the fall of '57 that he had committed suicide on McCall Glacier, where he was directing a glaciology project. (Photo courtesy Ed LaChapelle, no. 1340, 1952.)

FIGURE 230.
"Scientists drilling into a 3-meter wide crevasse to procure water for lubrication of a larger ice drill." (From a poster display; F-50-R681)

FIGURE 231.
Cal Heusser at one of the field camps. That is a sled made out of roofing material leaning on the tent to Cal's right. (Photo courtesy Ed LaChapelle, no. 585, 1952.)

FIGURE 232.
"Drilling to a depth of 450 meters 1½ kilometers from the main camp." (From a poster display; F-50-556.)

FIGURE 233.
We studied snow structure and density in snow pits. (F-50-R575)

FIGURE 234.
Ed LaChapelle weighing a core sample from a snow pit. Ed became a professor of glaciology at the University of Washington, specializing in avalanche research around the world as well as running the Blue Glacier Research Station in the Olympic Mountains. (Photo courtesy Ed LaChapelle, no. 566, 1952.)

FIGURE 235.
Mel Marcus (L) and Bucky Wilson (R) mapping the Taku Glacier with a theodolite. (F-50-R634)

FIGURE 236.
Duncan McCollester (L) and Robert Forbes (R), 1949. (AGS photo by Z. Stewart, S-49-359)

FIGURE 232

FIGURE 234

FIGURE 235

FIGURE 233

FIGURE 236

FIGURE 237A

FIGURE 237B

FIGURE 238

FIGURE 239

FIGURE 240

FIGURE 241

FIGURE 237.
In the beginning of the project, we used an Air Force SA 16 (Gruman Albatross, A), manned by the Tenth Air Rescue Squadron, to deliver supplies. We switched to C47s (B) when we discovered that later in the summer the snow was too soft and rutted for the SA 16 to take off. One time an SA 16 spent a week on the icefield waiting for snow conditions to improve. If the snow was especially sticky, we occasionally used JATO fuel (Jet Assisted Take Off) to help the C47s. That was a rather impressive display! (Both photos by WOF, 1950, no numbers.)

FIGURE 238.
The main camp on upper Taku Glacier in 1950. The tent in the center is the cook tent. The 10-meter mast at left is for the thermistor string to record temperatures at intervals above the glacier surface. (F-50-no number)

FIGURE 239.
The Columbia Icefield Chalet, built in 1938–1939, is located on the Banff-Jasper Highway (the Icefields Parkway) and overlooks the Athabaska Glacier. (F-53-K803)

FIGURE 240.
Jim Simpson, Sr. (L) with Jim Simpson, Jr. (R) (F-53-R480) at Southeast Lyell Glacier, Aug 8, 1953.

FIGURE 241.
Mount Columbia photographed by Jean Habel in 1901. (No number; copy photo from WOF Collection.)

FIGURE 242.

Mount Columbia and the
Columbia Glacier, Canada.
Photo by Mary T.S. Schäffer,
1907. (No number; copy
photo from WOF Collection.)

Research in the Canadian Rockies

In the late 1940s, I became interested in comparing the glaciers in a dry area with those in the wet coastal mountains that we had been looking at in the '20s, '30s, and '40s. The Canadian Rockies are 800 kilometers from the coast, and thus much drier. In addition to having a drier climate, the Canadian Rockies are one of the few places south of Alaska where glaciers terminate in the timber. Thus you have a method of dating which you don't have when the glaciers terminate far above timberline.

I visited the Canadian Rockies in 1948, '49, and '53.[16] There was a highway to the Columbia Icefield in Canada by 1940 that made access much easier (fig. 239). However, the only way to get to the Columbia Glacier still was by a three-day pack-train trip up the Athabaska River. That first year, '48, I hired young Jim Simpson as our packer to take us to the Columbia Glacier, and I came to know both him and his father well (fig. 240). The Columbia Glacier is very interesting. It had been visited back as early as 1901 by a German Jean Habel[17] (fig. 241) who, I believe, was the pioneer in the Athabaska Valley. The area was later visited by Mary Schäffer[18] in 1907 (fig. 242), B.W. Mitchell in 1912, B. Mitchell and H. Bryant in 1916, the Interprovincial Boundary Commission in 1919 (figs. 243, 244), J. Monroe Thorington[19] in 1923, and Byron Harmon, the famous photographer from Banff, in 1924 with Lewis Freeman (fig. 245).[20] I always admired Harmon's photography. While on my '24 trip to the Canadian Rockies, we passed Harmon and his pack train on their way to Jasper. After we parted, he took a magnificent picture of Mount Columbia. It's still a classic, clouds just breaking and the mountains clothed with snow (fig. 246).

FIGURE 242

Over the three visits to the Canadian Rockies, I reoccupied the positions of the earlier people and repeated their photography of the Saskatchewan and Athabaska glaciers, as well as my own from my first trip to the Canadian Rockies in 1922 (figs. 247–251).[21] I wanted to determine the amount of shrinkage that had taken place since the last set of observations. The history of photography in the Canadian Rockies goes back into the 1880s, so there is something comparable to that of the photographic record in Glacier Bay.[22] We found most all the glaciers receding—very few advancing. To strengthen the dating part of the study, Dr. Calvin J. Heusser, a botanist who was working with me on JIRP, came along in '53.[23] He took core samples from the muskeg to create pollen profiles from which to study the type of vegetation, and how it had changed over whatever length of time. He wanted to determine the post-glacial history of the area, which began with the end of the Cordilleran ice sheet.[24] The results of the 1953 trip were written up in the *Canadian Alpine Journal*[25] and for the meeting of the General Assembly of the International Association of Hydrology.[26]

Glaciation in the Northern Hemisphere

In the early '50s, the U.S. Army was becoming more and more interested in geography on a worldwide basis, including glacierized areas. They felt that while there was much information published about glaciers and glacierized areas, the material had not been summarized in a convenient reference work. Thus, I was contracted as principal investigator to "assemble this information in a form accessible to military users, and to supplement it with new information derived from the study of air photos and reports of expeditions."[27] This led to an effort by a number of people spanning several years and ended with publication of *A Geographic Study of Mountain Glaciation in the Northern Hemisphere,* a three-volume account of what was known of the glaciers in the temperate regions in the Northern Hemisphere.[28] We concentrated on those regions rather than the polar regions because of the availability of measurements of the critical changes that were occurring. This publication was considered state of the art at the dawn of the International Geophysical Year.

FIGURE 243

FIGURE 244

FIGURE 245

FIGURE 246

FIGURE 247

FIGURE 243.
Mount Columbia and
Columbia Glacier, Canada.
Picture taken by the
Interprovincial Boundary
Survey in 1919, from Warwick
Mountain, Sta. 109. (No
number; copy photo from
WOF Collection.)

FIGURE 244.
Castleguard Valley. Photo
taken by the boundary survey
in 1919. (No number; copy
photo from WOF Collection.)

FIGURE 245.
Byron Harmon (1876–1942)
was a famous photographer of
the Canadian Rockies region.
He spent the greater part of
his life exploring and hunting
in the Canadian Rockies. This
portrait was taken in 1924
by Lewis Freeman during the
Columbia Icefield Expedition.
(From the Byron Harmon
Collection, Whyte Museum of
the Canadian Rockies, Banff,
Alberta.)

FIGURE 246.
Byron Harmon's classic
picture of Mount Columbia,
Sep 25, 1924. This is a
postcard I bought in Banff.
(From the WOF Collection.)

FIGURE 247.
View of Saskatchewan Glacier
from Parker Ridge taken by
Mitchell in 1912 (top; copy
photo from WOF Collection);
my photo from Sta. 21 taken
in Sep 1922 (center; F-22-
153); and my photo from Sta.
21 in Aug 1948 (bottom;
F-48-206). There was a
recession of about 825 meters
from 1922 to 1948. Notice
also the change in the hanging
glacier in the upper center.

FIGURE 248.
Athabaska Glacier from the south ridge of Wilcox Mountain (A), Sep 1922 (F-22-CR-158c), and from nearly the same position (B), Aug 6, 1948 (F-48-R228). Note the Banff-Jasper Highway in the left foreground on the 1948 photo.

FIGURE 249.
Our vehicle in 1953 parked where one leaves the highway to climb Parker Ridge. This was one of the most efficient trips I ever made because we could drive all our equipment right to the photo/survey station or to the trail that led to the station. (F-53-R797)

FIGURE 250.
Robson Glacier in 1908 (A) taken by Rev. George B. Kinney, a Canadian mountaineer who made the first attempts to scale Mount Robson between 1907 and 1909 (no number; copy photo from the WOF Collection) and my photo (B) taken in 1953. (F-53-R298)

FIGURE 248A

FIGURE 248B

FIGURE 249

FIGURE 250A

FIGURE 250B

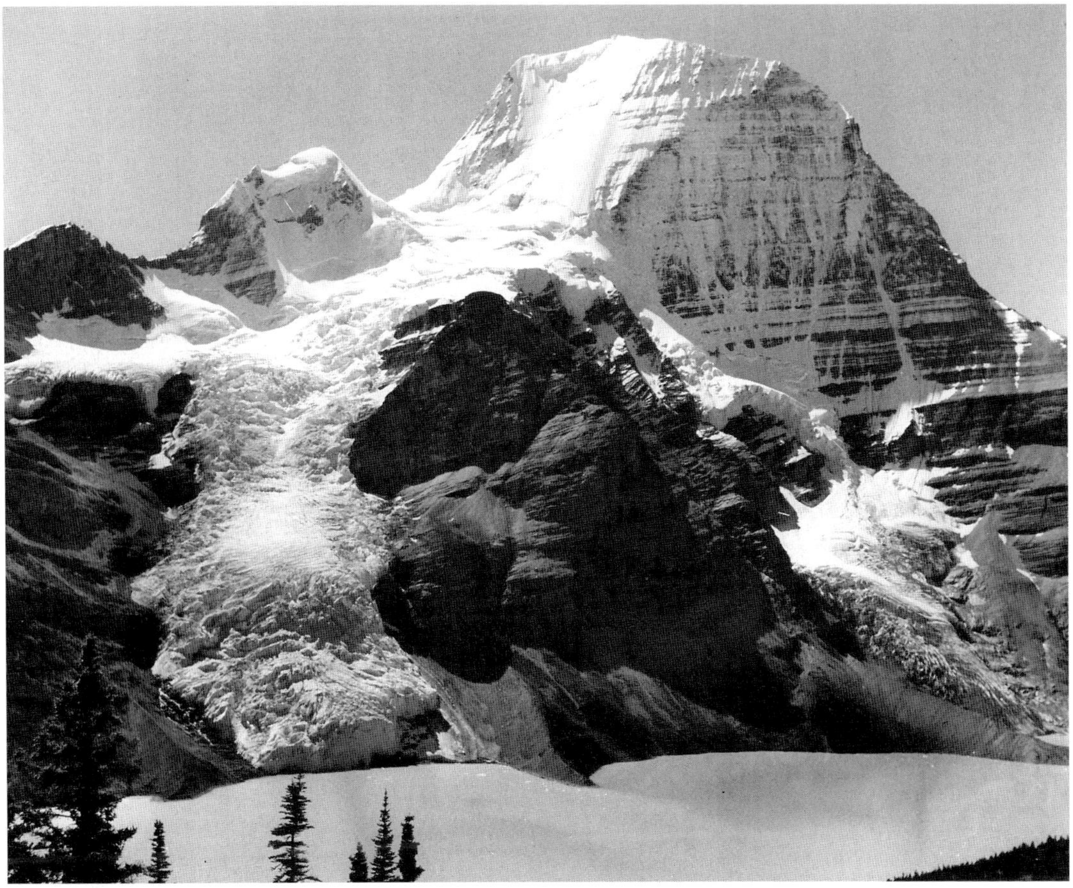

FIGURE 251

FIGURE 251.
Berg Lake, Berg Glacier and Mount Robson from Mumm Ridge. (F-53-R289)

Notes

1. The American Geographical Society (AGS) is an independent, not-for-profit corporation established in 1851 as a learned society to gather and disseminate geographical knowledge. It was the earliest such institution founded in America. (From an information paper published by the AGS in the 1950s.)

2. Sir John Franklin (1786–1847) pioneered English exploration in the arctic area. On his first arctic expedition in 1819, he attempted to determine the latitude and longitudes of the northern coast of North America. He led his second expedition to the Arctic in 1825 and 1826. In 1845, Franklin led the best-equipped expedition to enter the Arctic up to that time, with the ships provisioned for three years. He discovered a Northwest Passage, but he and his crew died during the expedition. When no one returned from the voyage, search efforts were begun in 1848, both public and private, English and American, and were unparalleled in maritime annals. The search resulted in a full exploration of the arctic region. In 1859 distinct traces of the lost expedition were found. (Stephen and Lee, *Dictionary of National Biography from Earliest Times to 1900*, n.d., 631–637.)

3. Wood, "The Parachuting of Expedition Supplies: An Experiment by the Wood Yukon Expedition of 1941" (1942, 36).

4. Wright, *Geography in the Making: The American Geographical Society, 1852–1952* (1952). See also Wright, "The American Geographical Society: 1852–1952" (1952).

5. See Appendix C for a summary of the trip.

6. Once Bill officially became part of the scientific community, he has adopted the metric system of measurement.

7. Field, W. O., "Glacier Recession in Muir Inlet, Glacier Bay, Alaska" (1947).

8. See Chapter 3, note 14 for references concerning the 1931 Mount Fairweather climb.

9. Frank Heintzleman (1888–1963), a forester, was territorial governor of Alaska from 1953 to 1957. He previously served as regional forester for the territory (1937–1953) directing the work of the U.S. Department of Agriculture's six bureaus in Alaska and

served from 1941 to 1953 as commissioner. He was one of the foremost authorities on the history, the people and the resources of Alaska. (Atwood and DeArmond, *Who's Who in Alaskan Politics*, 1977.)

10. Wright, "The American Geographical Society: 1852–1952" (1952).

11. Dr. Thomas Poulter's life spanned more than half a century of research and exploration. In 1933 he joined the Byrd Antarctic Expedition as second in command and led the rescue party that saved Admiral Byrd's life. He later designed the Antarctic Snow Cruiser used on the U.S. Antarctic Service Expedition. He spent thirty years at Stanford Research Institute (SRI), becoming associate director of the newly formed SRI in 1948 and founding SRI's Poulter Laboratory in 1953. During this period he went to Alaska to make seismic measurements on Taku Glacier. (*Spectrum*, July–August 1978.)

12. See Appendix E for a summary of Hubley's accomplishments.

13. Mel Marcus died March 2, 1997.

14. During the years the project was run by the American Geographical Society, JIRP stood for Juneau Ice Field Research Project. After the AGS terminated its involvement in the project in the mid-50s, it was taken over by Maynard Miller with the acronym JIRP standing for Juneau Ice Field Research Program. See Bohn, "The Juneau Ice Field Research Project" (1958).

15. A list of JIRP reports issued by the Department of Exploration and Field Research of the American Geographical Society is in Appendix E.

16. See Appendix C for a summary of the trips.

17. See *Appalachia* (1902).

18. For a description of Mary Schäffer's trip, see her article in the *Canadian Alpine Journal* and her book *Old Indian Trails* (1911).

19. Thorington, *The Glittering Mountains of Canada* (1925).

20. Freeman, *On the Roof of the Rockies* (1925).

21. See Field, W.O., "Glacier Observations in the Canadian Rockies" (1949).

22. See Field, W.O., "Glacier Observations in the Canadian Rockies" (1949); Meek, "Glacier Observations on the Canadian Cordillera" (1948); and Wheeler, *Canadian Alpine Journal* (1931).

23. Calvin J. Heusser (1924–) received his Ph.D. in botany in 1952. In 1953 he became a research associate at the American Geographic Society. During his fourteen years with the society he was administrator of the Juneau Ice Field Research Project from 1953 to 1959. Dr. Heusser was an associate professor in the departments of biology and geology at New York University from 1967 to 1971, and in 1971 became a professor of biology, a position he still holds, specializing in palynology.

24. The most recent ice age in North America, the Wisconsin, began about 120,000 years age, reaching an extreme about 70,000 years ago. Then the climate moderated slightly and the ice sheets began a slow retreat but the climate still remained severe. By 35,000 years ago the ice sheets were spreading out once more, reaching their greatest advance 18,000 years ago at the bitterest stage of the ice age. The climate then moderated and an interglacial period began about 10,000 years ago, which continues to this day. In North America, the ice sheet centered over Hudson Bay was called the Laurentide ice sheet and covered most of eastern Canada and extended deep into New England and much of the Midwest. The Cordilleran ice sheet, centered over the Canadian Rocky Mountains, covered parts of Alaska, much of western Canada, and parts of Washington, Idaho, and Montana. About 15,000 to 18,000 years ago the ice sheets began to retreat and by 7,000 years ago they had all but ceased to exist. (Imbrie and Imbrie, *Ice Ages*, 1979; Coch and Ludman, *Physical Geology*, 1991; and Chorlton et al. eds., *Planet Earth: Ice Ages*, 1983, 17–30.)

25. Field and Heusser, "Glacier and Botanical Studies in the Canadian Rockies, 1953" (1954).

26. Heusser, *Glacier Fluctuations in the Canadian Rockies* (1951, 493–497).

27. Field, W.O. et al., *Geographic Study of Mountain Glaciation in the Northern Hemisphere, Pt. 10: Atlas of Mountain Glaciers in the Northern Hemisphere* (1958, Technical Report EP–92, iii [this was a separate publication from the text]).

28. Field, W.O. et al., *Geographic Study of Mountain Glaciation in the Northern Hemisphere* (1958).

FIGURE 252

CHAPTER 8
AN INTERNATIONAL SCIENTIFIC PERSPECTIVE
1956–1963

The International Geophysical Year was the idea of Lloyd Berkner in the spring of 1950.[1] The planning stage spanned the early 1950s, and the actual fieldwork began in July of '57 and lasted until the end of '58. The whole concept of the IGY was very different from anything we'd had before. This was the first time there was going to be a coordinated program of observations all over the world. This was in contrast to the International Polar Years of 1881–1882 and 1931–1932. Those efforts were concerned primarily with studies unique to the polar regions, such as the aurora, cosmic rays, and gravity in the polar regions, and included just the few countries that were involved in those studies. On the contrary, the IGY included studies worldwide, involving fifty-four nations, and it provided a chance for people to travel to other countries to learn what the programs were, and why and how they were carried out. The political climate in the '50s made this a good time for such international exchange. So for the first time, we could meet and become friends, as well as

colleagues. And I think this was one of the most important aspects of the IGY. It was hoped the IGY would be the beginning of a coordinated effort in science, and to a great extent that occurred. It opened a new era, and I think in the next century we'll see a return to such programs. Human existence on earth depends on an intelligent appraisal of the resources we have worldwide.

Originally, there was to be no glaciology. It was added well after the original plans were made. Somebody mentioned that because glaciers were also changing features of the earth's surface, they should be included.[2] From the global standpoint, it is very important to know whether glaciers are receding or advancing, thickening or thinning. First, if the ice isn't on the land surface, then it's in the ocean. Dr. Mark Meier has since calculated[3] that sea level could rise 0.08 to 0.25 meters in the next 100 years—because of ice melt—if the air temperature rises 1.5° to 4.5° C. Certain changes in the glaciers are also very good indicators of long-term climate change.

FIGURE 252
A U.S. Navy plane equipped with skis preparing for takeoff from an airstrip on the Ross Ice Shelf. (ANT-F-57-536)

87

I became involved in preparations for the IGY about 1954. I received good support from the AGS during the whole IGY period. The director recognized the importance of such an international cooperative effort. The IGY was a very lively period for me. There were lots of trips to Washington, D.C. for meetings, as well as to Switzerland, Sweden, Russia, and other countries. I met some extremely interesting and imaginative people, and whether or not they'd ever heard of a glacier before, they were immediately enthusiastic about studying this feature of the earth. The chairman of the National Committee for the IGY was Joseph Kaplan from UCLA [University of California at Los Angeles]. After one of our early conferences he said, "Field, I think glaciology is going to be a sleeper. I think it's going to have a lot of support and be very interesting." I don't think he'd ever considered it before. Allen Shapley, who was one of the IGY vice presidents, was interested, as was Hugh Odishaw, who was the executive secretary of the U.S. National Committee in Washington. I got to know him well. Pem Hart at the Academy of Science was also one of those people. He is still at the Academy. I continued to work with many of these people in the years following the IGY.

I was appointed to the Panel for Glaciology and subsequently became chairman of it. Glaciology is concerned with hydrology, meteorology, physics and geography, and thus requires a multidisciplinary approach. So the members on the Panel for Glaciology were involved in all kinds of research and different approaches to the study of ice and snow on earth.[4] I learned a great deal from these associations. I wouldn't have missed the IGY and post-IGY period for anything.

I was also the consultant on the Antarctic Committee.[5] This sort of mystified me, because I hadn't been to the Antarctic, but it turned out there were very few people who had been there. However, the IGY chairmen wanted people on the committee who were involved in related studies, such as the glaciology program, so that was why I was selected.

I was invited to go to the Antarctic in the fall of '57. This was a great experience.[6] I'd never been there nor ever expected to go. Suddenly, there I was heading for the Antarctic. Our trip was quick, but we saw a great deal (map 40; figs. 252–254). We landed at McMurdo Station, the main naval base and airfield between New Zealand and the Antarctic. A few days later we flew to Little America 4 and Little America 3. Little America 3 was formed just before World War II and was abandoned when the war broke out. We found the site of the pit study,[7] with some of the equipment buried in snow. We also had a day

trip to Byrd Station, 800–1000 kilometers away. On the two or three flights around McMurdo Sound, we could see the Wright Valley that is one of several dry valleys. It is one of the most interesting features in Antarctica, a valley with glaciers coming down both sides but most of the valley free of ice. Fortunately we also had a flight over the South Pole. We circled the pole several times and I got a good photograph looking down on it. Sir Hubert Wilkins[8] was also in Antarctica with us. He was an old-timer and a very pleasant companion. Sometimes he would join us on a flight and just having Sir Hubert with us made us feel "here's somebody who knows, somebody who's been around for awhile."

We returned to Newport News, Virginia, on a cold night around the fifteenth of December. A few of us took a bus to Washington and had to stand up all the way, a several-hour trip. It was snowing and cold and blowing, and most of our heavy clothes used in the Antarctic were packed up in duffel bags. I think that was the coldest we ever were on the whole trip.

Alaska During the IGY

During the IGY, the AGS received support for three trips to Alaska to continue the study of the fluctuations of glaciers. We tried to visit all the glaciers we had studied before, as well as glaciers in some key areas not previously covered. The rationale for using IGY funding was to establish a database from which we could determine what had gone on before, as well as use as a benchmark for future observations. IGY funding allowed us to take people from different disciplines besides glaciology, such as botany, palynology, and forestry, to name a few. We also were able to have larger field parties, so we could work faster, cover more area, and take more measurements.

The first year, '57, a party assembled to work in Prince William Sound, on the Kenai Peninsula, and in the Wrangell Mountains in interior Alaska.[9] There was also a brief unplanned trip to Mount McKinley to document a surge of the Muldrow Glacier that had just occurred. The next year, '58, the effort in Alaska was to visit all the glaciers that we had visited on previous trips, primarily in southeastern Alaska.[10] On July 11, before I left for Alaska to join the field party, there was a severe earthquake along the Fairweather Fault at the southern end of the Fairweather Range. The fault goes along the west side of the Fairweather Range, and ends up in Nunatak Fiord at the head of Yakutat Bay (map 39). Many of us were interested in investigating what effects,[11] if any, earthquakes have on glaciers, such as a great discharge of ice from the tidewater glaciers along the coast. Within two days of the earthquake Ken Loken, a pilot in Juneau, took some of the field party on a flight over Lituya Bay, Glacier Bay, and the surrounding area but they found little or no change in the glaciers, except a big change at Lituya Glacier. The T at the head of Lituya Bay is on the Fairweather Fault. A huge slide at the head of this over-steepened fiord came down and sheared off the lower 300 meters or so of Lituya Glacier, and sent up a tremendous wave that washed all the way out to the entrance of the bay, stripping the vegetation as high as 550 meters on the north side of the bay (figs. 255, 256).[12] At the time this slide occurred, two fishing boats were anchored in the lower end of the bay, on both sides of the entrance. One fisherman saw this wave coming and began to pull anchor. The wave caught his boat, and swung it right over the spit and into the open ocean. The man lived to tell the tale.[13] It was an amazing escape. The other boat was swamped and it and everybody on board were never seen again. These waves had happened before. Presumably in the '30s there had been a big wave. Until the wave in '58, no one could explain what caused them.

I joined up with the field party and we proceeded to take a trip around Glacier Bay in the usual way. We had access to a National Park Service boat that summer, the *Nunatak* (fig. 257). Aside from Glacier Bay, we did quite a lot in the Coast Mountains, and we continued to watch for any effects of the earthquake on the glaciers. When I returned to Glacier Bay and Prince William Sound in '61 there was still no evidence of any change in the glaciers due to the earthquake.[14]

FIGURE 253

FIGURE 254

FIGURE 255

FIGURE 253.
Byrd Station, a U.S. scientific outpost at latitude 80° south, longitude 120° west on the inland ice. Its elevation is about 5000 feet [1700 meters] and soundings indicate about 10,000 feet [3000 meters] of ice beneath. (ANT-F-57-543)

FIGURE 254.
View out from a snow cave in the Ross Ice Shelf near the U.S. IGY station at Kainan Bay. The floor of the cave is bay ice which goes out during the summer months. (ANT-F-57-629)

FIGURE 255.
The large rockslide of July 9, 1958 sheared off part of the front of Lituya Glacier and caused water to surge over the spur opposite.

FIGURE 256.
A giant wave generated by a rockslide from the cliff (r) at the head of the bay destroyed the forest over the light areas to a maximum altitude of 1,720 feet [575 meters] at d and to a maximum distance of 3,600 feet [1,200 meters] in from the high-tide shoreline at Fist Lake (F). A fishing boat anchored in the cove at b was carried over the spit in the foreground; a boat under way near the entrance was sunk and a third boat, anchored at e rode out the wave.

FIGURE 256

FIGURE 257

FIGURE 257
The M/V Nunatak II *in Muir Inlet, Sep 1, 1958. (F-58-K353)*

FIGURE 258
Hubbard Glacier in Yakutat Bay. Photo taken from near the summit of Haenke Island by I.C. Russell (L) in 1890 or '91 (Russell photo no. 390; copy photo from the WOF Collection) and by the AGS in 1959 (R). (M-59-P137, 138)

We returned to Alaska in '59 primarily to visit the glaciers along the lower Copper River. Yakutat Bay, Disenchantment Bay, and Russell Fiord were the big gaps in our area of coverage (map 41).[15] None of us had been to these areas before to reoccupy the early boundary survey stations[16] as well as the stations that I. C. Russell[17] had occupied dating from 1891. This was a very interesting area worth considerable attention then, because Hubbard Glacier, one of the biggest glaciers in Alaska, was advancing at a rather fast rate, perhaps as much as fifty meters a year. It had advanced considerably since Russell's visit in 1891 (fig. 258), the Harriman Alaska Expedition of 1899, the visit of Tarr and Martin in 1909 and 1910 and the survey by members of the Harvard Mountaineering Club in 1946.[18] In 1959 the high trim line in Russell Fiord, formed presumably back in the eighteenth century when Hubbard Glacier moved forward to Gilbert Point, sealed off the entrance to Russell Fiord and formed a lake, was still visible. As I recall the lake rose up about sixty meters before the lake drained out past Hubbard Glacier. This is what happened in 1986 when the glacier again advanced, crossed the fiord and dammed it up.

I didn't think I would ever get to this area because of the difficulty of arranging transportation from Yakutat. It's seventy or eighty kilometers to the Hubbard Glacier from Yakutat, and much further up to the head of the fiord. You travel up the big bay, and the swells from the Gulf [of Alaska], which come rolling into the bay, can make it pretty rough at times. There are very few vessels that are available for charter, but we managed to arrange for one. When we arrived at Yakutat, however, there was considerable delay in starting. It turned out that the boat had sunk the week before and it took several days to clean it out and prepare it for our trip. It was not really ready by the time we took off. We ran on batteries because the generator was inoperable. The owner of the boat had flown down to Juneau to get a new generator and he'd crash-landed on the beach on Icy Strait on the way back. About halfway through the trip he returned in an outboard motor boat with the new generator, and inserted it. However, it was one of those generators that went on the opposite way from every other generator and we couldn't use it. This made quite a delay. I think it was the funniest trip I remember. The harder things got, the funnier everything seemed to be. Everybody was in a very jolly mood most of the time. So that '59 trip to Alaska ended what was considered the IGY period and we tend to date things from those three years of observations.

"Don't Forget the Little Glaciers"

One project during the IGY was to map in detail a few relatively small glaciers without extensive expeditionary effort. This project produced a volume published by the AGS called *Nine Glacier Maps.*[19] The intent was to repeat this mapping in five, ten or fifteen years to determine the volume change of the glaciers in that period, and then relate it to climatic change. Mapping the terminus is a quick way of determining the current health of a glacier, but it does not take into account the time lag that occurs between the beginning of a climatic change and when the effect of that change shows at the glacier terminus. That is

FIGURE 257

FIGURE 258A

FIGURE 258B

especially true of the big glaciers. The study of small glaciers began in this country in the early '50s. Hans Ahlman, visiting the Juneau Icefield in '52, said there was information to be gained from the icefield, but because it was such a big mass of ice it would be very difficult to do detailed studies that would be meaningful in determining the effects of climate on glaciers.[20] He suggested studying a glacier only a few kilometers long and thus, with fewer logistical problems, one could get more information. Also, on a smaller glacier any climatic effects would be apparent sooner, and would be easier to discern. We continued with the Juneau Ice Field Research Project to determine features of these big ice masses, but we also put some effort into the Lemon Creek Glacier, a small glacier just above Juneau. Thus, the effort shifted, but not exclusively, from looking at the big, spectacular glaciers to the smaller glaciers. If somebody wants me to suggest a program, certainly look at the big glaciers, but don't forget the little ones. This is also true as you go back over the old pictures. It used to be you took pictures of the big glaciers, such as the Muir, and sometimes in the background there was a little glacier, and that was the one that changed but no one noticed it. They were looking at the big glaciers.

IGY and the World Data Center A for Glaciology

After my trips to Alaska in '26 and '31, I discovered there was no one inclusive collection of glacier pictures. So I saw the need to assemble as many photographs as possible taken by people on earlier expeditions, photographs taken by such groups as the Harriman Expedition.[21] And a big collection of photographs taken by members of the Geological Survey in their rambles through Alaska. These had not been published because the Survey's primary objective was exploration of mineral resources, not glacier observations. Another valuable collection was from the boundary survey efforts.[22] Then the navy air photography began in '29 in southeastern Alaska. Those pictures resulted in the first aerial maps of Glacier Bay, and then in '34 the navy took some oblique photos. In '41 the U.S. Army Air Corps made trimet-mapping flights because the pilots flying war-related missions needed to know where the mountains were.[23] These quick aerial navigation charts at about 1:100,000 were made showing most of the topography. These charts primarily covered the air routes from Edmonton to Anchorage and Edmonton to Fairbanks. This was a big shot in the arm for aerial photography.

And then in '48 southeast Alaska was mapped very precisely by the navy with vertical photography. The Defense Department wanted to map all of Alaska. They began with southeastern Alaska in '48 and continued up the coast in later years. In both '29 and '48, Martin,[24] as chief of the Division of Maps at the Library of Congress, as well as a colonel in the army, asked the navy, since they had all those people up in the air, if they would take some oblique pictures of glaciers. And they did. The first aerial pictures taken by a glaciological team was in '31. After Wright and Reid finished working in Glacier Bay that summer,[25] they took a flight over southeastern Alaska with the photographer Percy Pond. We did not know until recently that Pond was on that trip, so any 4 by 6 inch aerial photos of southeastern taken in '31 were probably those taken by Pond.

As I began to put together this collection of photographs, I discussed with my connections in Washington the possibility of starting a collection that would be for my projects as well as for the general interest of glaciologists in this country. In May of '32, there was $100 left from Wentworth and Ray's trip to Alaska for the Geological Survey and, if not spent, it would revert to the U.S. Treasury on July 1.[26] Some of us, including Lawrence Martin, François Matthes,[27] and Will Wright[28] discussed the possibility of starting a collection of pictures with that hundred dollars. Martin agreed to store such a collection in his office at the Library of Congress. Word of the collection spread and C. Hart Merriam, the editor of the publication covering the 1899 Alaska Harriman Expedition, said he had some extra prints I could have. Part of the hundred dollars went into a cabinet, which I still have. The collection was directed and financed from then on through the Research Committee on Glaciers of the American Geophysical Union, of which François Matthes was chairman. Thus, it became the AGU Collection. We kept adding any pictures, maps, and documents related to the current study of glaciers, primarily in the United States. When Martin retired from his position as chief of Division of Maps in the middle '40s, Matthes suggested that the collection be sent to the American Geographical Society because it had an actively operating program of glaciological studies. So I agreed to keep the American Geophysical Union Collection, and it became known as the AGU/AGS collection.

H.F. Reid, with whom I had become good friends, died in 1945 and his wife gave me his enormous collection of 6.5-inch by 8-inch glass plates, including ones he'd taken in Glacier Bay in 1890 and '92.[29] So I drove down to Baltimore

FIGURE 259

Pem Hart (L), Robert Haefeli and Barclay Kamb (R) at the Davos research station. Hart was with the U.S. National Academy of Sciences / National Research Council and was the staff officer assisting the Committee on Polar Research in organizing programs, especially glaciology, for the IGY and continued in that capacity after the IGY. Haefeli was a Swiss research engineer at the Davos station. Kamb was a visiting U.S. researcher who later became a professor at the California Institute of Technology. (F-60-S3)

FIGURE 260

Marcel de Quervain, director of the Weissfluhjoch in Davos when I was there in the '60s. (F-67-E219)

FIGURE 261

Herfried Hoinkes (L) from the University of Innsbruck and Mischa Plam (R), director of the Problem Laboratory on Snow Avalanches and Debris Flows at Moscow State University, 1971. Plam later left Russia and became director of the Institute of Arctic & Alpine Research's Mountain Research Station in the mountains west of Boulder, CO. (Photo courtesy M. Meier.)

FIGURE 262

V. Kotlyakov (L), Gregor Avsiuk and Mark Meier (R), 1970. Kotlyakov was a glaciologist with the Institute of Geography USSR Academy of Sciences and later became president of the International Commission on Snow and Ice, attaining the rank of academician. Avsyuk ran the entire program for glaciology in the USSR and was the chief person in the IGY for the USSR. (Photo courtesy M. Meier)

FIGURE 259

FIGURE 260

FIGURE 261

FIGURE 262

in a station wagon, picked up all these boxes of glass plates, and drove them back to New York. What I remember most vividly were the many cobblestone streets in Baltimore. I had to go very gingerly over these. I then had prints made of those in which I was interested. Thus, I collected quite an enormous number of photos, and my office became a center for the collection of these.

During the IGY, three centers were selected for the collection and storage of all the scientific data and photographs being gathered. Because of the huge AGU/AGS photo collection and all the activities we were doing at the AGS, we were awarded the IGY World Data Center A for Glaciology in North America in 1957. This remained the situation until 1970 when the Data Center, including the photos, was transferred to Project Office Glaciology of the U.S. Geological Survey in Tacoma, Washington. That office already had all of Austin Post's photographs, the most complete aerial coverage of glaciers in Alaska and the lower 48 since 1960.[30] I retained the material that I had acquired for my personal research. The USGS maintained the Data Center until it was transferred in 1976 to the Institute of Arctic and Alpine Research in Boulder, Colorado, where it is now.

An International Scientific Perspective

For me, the aftermath of the IGY was the most significant part, as it developed an international scientific perspective that continues to this day. We saw common problems relating to snow and ice all over the world, and this made a big difference in our outlook. IGY activities allowed us to compare the behavior of glaciers in the Southern Hemisphere with those in the Northern Hemisphere. The IGY also helped to make glaciers a recognized part of the environment and a respected study that warranted continuing support, similar to those in Scandinavia and Switzerland, where glaciological study programs were initially developed and are continually supported.

I became friends with scientists from all over the world (figs. 259–265). On one meeting in Lausanne, Switzerland, we were honored to be met at the railroad station by the eminent scientist Louis Mercanton. He was a very old man by then. He had begun studying the Rhône Glacier in the 1870s and had produced a magnificent monograph describing the very careful mapping of the glacier year to year.[31] He was one of the great scientists of the nineteenth and twentieth centuries.

In Switzerland, besides Mercanton, I met Peter Kasser. He was an engineer at the Swiss Federal Institute of Technology (ETH) in Zurich and was taking ice cores from glaciers. Kasser was a kind person and very enthusiastic. During a glaciological meeting in Switzerland in 1961, we also visited the Weissfluhoch—the snow, ice, and avalanche station in Davos located in the Engadine area in Switzerland. The facility was supported primarily by the Swiss Glaciological Commission and had many full-time scientists as well as guests from different countries. Herfried Hoinkes from the University of Innsbruck was spending a year there studying the relationship between glaciers and climate. I first met him when I was in the Antarctic in '57. He was a delightful and very learned man. Unfortunately, he died very suddenly a few years later. I recall one afternoon showing him as many of the sights of New York as possible, and I always wondered how much he understood.

We also got to know the Russians pretty well, such as Shumskiy, Tushinsky, and Avsiuk. The Russians have quite a history of glacier studies and they are very serious about mountain research. When the Germans invaded Russia in 1941, they were using maps dated about 1875, that showed a glacier on Mt. Elbrus level with the top of a pass, and the Germans thought they could come right across the pass, tanks and all. However, by 1941 the glacier had thinned and was 100 to 150 meters below the pass. So there is even application of glacier studies to military intelligence and logistics! In the Alps, the Swiss have been studying the glaciers meticulously since the early 1800s. The glaciers are much more closely related to the economy of that country—and others around the world—than they are in the U.S. In central Asia, most of the water comes from the melting of snow in the Pamir Mountains, and in northwestern China the ground water tapped by wells is largely fed by meltwater from snow and glaciers in surrounding mountains. So if the snowline rises and the small glaciers or snow patches disappear, it can affect the whole economy of the region. This is also true in parts of California, the Pacific Northwest and parts of Alaska. So we were influenced by these experienced scientists from abroad, who were brought up in areas with glaciers. The study of glaciers in such countries is not something one picks up as an exotic way to get to the mountains, but one upon which their economy depends.

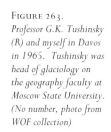

FIGURE 263

FIGURE 263.
Professor G.K. Tushinsky (R) and myself in Davos in 1965. Tushinsky was head of glaciology on the geography faculty at Moscow State University. (No number, photo from WOF collection)

FIGURE 264.
In September 1961 there was a glaciological meeting in Seattle, Washington. We took a day trip to Mount Rainier, only two hours away. Many of the glaciologists I had come to know abroad as well as the growing number of glaciologists from this country were there, including Dr. Mark Meier (in back with camera) and Austin Post (center) from the USGS Project Office Glaciology in Tacoma, WA. Mark was the Project Chief, a position he held until his retirement in the early '80s. I am in the foreground. (No number)

FIGURE 265.
Mark Meier (R) and myself at Mount Rainier, Washington, 1961. (No number, photo from WOF collection)

FIGURE 264

FIGURE 265

FIGURE 266.
Our wedding photograph, 1963. (No number)

FIGURE 267.
Talking with Jim Simpson, Sr., in front of his Num-Ti-Jah Lodge in 1963. (MLF-63-3-3)

FIGURE 268.
In 1989 we returned to Num-Ti-Jah Lodge, our first visit since our honeymoon in 1963. (Photo by C S. Brown)

My New Family

In March of 1960, my first wife Alice died very suddenly of pancreatitis. Fortunately, all the post-IGY activities kept me busy after that. Then, during the numerous trips to Europe after the IGY for meetings of the International Glaciological Society and the International Commission of Snow and Ice, I renewed acquaintances with an old friend, Mary Mapes, who was living in Geneva, Switzerland while working for the World Health Organization (WHO). Around Easter in 1962, Mary became the WHO representative for information to the United Nations in New York and we decided to get married a year later, on the eighth of August, 1963 (fig. 266). After the ceremony, we flew to California for a meeting I was to attend at UCLA, so we spent the first part of our honeymoon quartered in the women's dormitory there! Then we continued to the Canadian Rockies and stayed at Num-Ti-Jah Lodge at Bow Lake, built and operated by my friend Jim Simpson (figs. 267–268).

FIGURE 266

FIGURE 267

FIGURE 268

Notes

1. The idea for an IGY in 1957–1958 was first suggested in April 1950, at an informal gathering of scientists in Silver Spring, Maryland, by Dr. Lloyd V. Berkner, one of the world's foremost ionospheric scientists. The International Council of Scientific Unions agreed to the proposal and invited representatives of the national scientific organizations of the world to a meeting to discuss the program and problems. It was only after this meeting, in 1954, that the governments of the 64 nations and private institutions were asked to join and lend their support. (Marshack, *The World in Space: The Story of the International Geophysical Year*, 1958, 11).

2. The U.S. IGY program in northern latitudes included observations of glaciers and sea ice on a more extensive and comprehensive scale than had previously been attempted. In Alaska, detailed studies of glacier regimen were carried out at three localities, eight glaciers were mapped, and the condition and behavior of the termini of many other glaciers were determined by ground and air reconnaissance. Detailed studies were made of the structure, flow, and mass budget of two glaciers in Washington State; less detailed studies were made on six other glaciers in Washington, Oregon, and Montana. (A.P. Crary, W.O. Field, and Mark F. Meier, "The United States Glaciological Researches During the International Geophysical Year," 1962, 12).

3. Meier, "Contribution of Small Glaciers to Global Sea Level" (1984, 1418).

4. Members of the Panel for Glaciology:
   Henri Bader, SIPRE; University of Miami, Ohio
   Albert P. Crary, AFCRC; Arctic Institute of North America
   Phil E. Church, University of Washington
   William O. Field, American Geographical Society, Chairman
   Richard F. Flint, Yale University
   Richard P. Goldthwait, Ohio State University
   Richard C. Hubley, University of Washington
   Mark F. Meier, USGS
   Frank Press, Lamont Geological Observatory
   John C. Reed, USGS; Arctic Institute of North America

   George P. Rigsby, SIPRE; U.S. Navy Electronics Lab; Arctic Institute of North America
   Robert P. Sharp, California Institute of Technology
   Paul A. Siple, Army Research Office
   A. Lincoln Washburn, Dartmouth College; Yale University

5. The following summary of the proposed IGY Antarctic research shows how new a frontier the Antarctic was as recently as 1957:

   After much preliminary planning by its Technical Panel for Glaciology, the U.S. National Committee for the IGY proposed three basic principles to guide the United States' approach to Antarctic glaciological research: (1) investigations should concentrate on items which are peculiar to the Antarctic and which cannot be studied efficiently and effectively in more accessible areas; (2) attention should be given . . . to basic principles and to matters of world-wide significance; (3) efforts should be made to learn as much as possible about the physical state, environment, and behavior of Antarctic ice bodies. To satisfy these principles, certain lines of investigation were recommended: (1) measurement of ice thickness leading to a reliable calculation of the volume of Antarctic ice primarily by the seismic-reflection procedure and gravimetric surveys; (2) observations of variations in the volume of Antarctic ice in the past and measurement of current rates of change; (3) deep core drilling as a means of obtaining data at depth in the inland ice and the Ross Ice Shelf; (4) determination of firn stratigraphy; (5) studies of thermal regime; (6) studies of crystal fabrics; (7) measurements of glacier movement; (8) measurements of rates of accumulation and wastage to determine Antarctic glacier regime; (9) micrometeorological studies; (10) determination of climatic fluctuations as recorded by changes in the size of Antarctic glaciers (A.P. Crary, W.O. Field, and Mark F. Meier, "The United States Glaciological Researches," 1962, 6).

6. Bill wrote an article "Flight to the South Pole" about his trip to the Antarctic for the newspaper of the school his son John attended. It was published in *Quad*.

7. A general method of study of the characteristics of a snow pack is the pit study. A pit is dug in the snow at a given site at periodic intervals and characteristics of the snow, such as temperature, density, grain size, grain type, shear strength, ram resistance, cone hardness, tensile strength, and air permeability are recorded. When the records for successive pits at a given site are compared, they afford an insight into the changes taking place within the snow pack.

8. Sir Hubert Wilkins (1888–1958) was an Australian explorer, naturalist, and pioneer of polar aviation. He began working in the electrical business, then joined the film-making industry. This allowed him to do what he wanted to do, namely travel the world and see every country. He visited fifty-two countries in two years. He went to the Arctic with Stefansson in 1913–17 for his first taste of polar exploration. After that, polar exploration occupied thirty-two winter seasons, twenty-four in the north and eight in the south. In 1928 he and Carl Eielson became the first to fly an airplane across the Arctic Ocean from Point Barrow, Alaska to Spitsbergen, 3540 kilometers (2200 miles). Later in 1928 while surveying in the Antarctic he made the first Antarctic airplane flights. From 1942 until his death on December 3, 1958, Sir Hubert, still an Australian citizen, was attached to the U.S. Department of the Army as a civilian consultant. (WOF Collection, Office Files series, Clippings folder.)

9. See Appendix C for a summary of the trip.

10. See Appendix C for a summary of the trip.

11. In the decade following the 1899 earthquake, a series of glacial surges were observed which led Ralph S. Tarr and Lawrence Martin to propose a theory that such surges are a delayed response to earthquakes. That theory is known as the Tarr-Martin theory of earthquake advance. See Tarr and Martin, *Alaskan Glacier Studies of the National Geographic Society in The Yakutat Bay, Prince William Sound and Lower Copper River Regions* (1914, 168–197). See also Field, W.O., "The Effect of Previous Earthquakes on Glaciers" (1968, 252–260).

12. Miller, D.J., *Giant Waves in Lituya Bay, Alaska* (1960).

13. See Appendix E for an excerpt of a letter (courtesy of Austin Post, 1978, personal communication) from Mrs. John Turner, owner of the *Cameo*, possibly to her sister and brother-in-law. It gives the reader an idea of what one eyewitness thought of the earthquake on July 9, 1958, and the big wave in Lituya Bay.

14. See Appendix C for a summary of the trip.

15. See Appendix C for a summary of the trip.

16. Hubbard Glacier was first described by I.C. Russell after his visit there in 1890 for the USGS (see note 17 below). Russell was the first to photograph Hubbard on his trips there in 1890 and 1891, followed in 1895 by A.J. Brabazon, a Canadian working with the Alaska Boundary Commission (see Appendix E for a brief history of the boundary survey work), H. G. Bryant in 1897 while on an expedition to climb Mount St. Elias sponsored by the Philadelphia Geographic Society, G.K. Gilbert with the Harriman Expedition in 1899 (see Chapter 2, note 8), Tarr and Martin in 1905–1906 and 1909–1910 (see Chapter 3, note 22) and in 1911 by N.J. Ogilvie of the International Boundary Survey.

17. Israel Cook Russell (1852–1906) was a geologist, first with the Wheeler (Canadian/U.S. Border) Survey (west of the 100th meridian) in 1878, then with the U.S. Geological Survey. In 1892 he became professor of geology at the University of Michigan. He very nearly succeeded in climbing Mount St. Elias in 1890, but was prevented by storms (see "The Expedition of the National Geographic Society and the U.S. Geological Survey (1890)," 1891, 872). Mount Russell, Russell Fiord, Russell Glacier and Russell Island are named for him. Russell Island was named by Field and Cooper in 1937.

18. See Ferris, "Mount St. Elias" (1947); Miller, M.M., "Yahtsetesha" (1947); and Molenaar, "St. Elias: The First American Ascent" (1947).

19. American Geographical Society, *Nine Glacier Maps, Northwestern North America* (1960).

20. Hans Ahlman was born in Sweden in 1889. In 1929 he was appointed professor and director of the geography department at Stockholm University. Beginning in 1918, he worked on glaciers and their relationship to climatic fluctuations. Between 1931 and 1938 he led various expeditions to Nordaustlandet,

Spitsbergen and Vatnajokull. The creation of the Norwegian-British-Swedish Antarctic Expedition of 1949–1952 was a result of Hans' enthusiasm and initiative. He was chairman of the Swedish Organizing Committee for the expedition and was responsible for the type of glaciological program that the expedition carried out. From 1950 to 1956 Ahlmann was Swedish ambassador to Norway, but he still continued his publication of glaciological articles. (*ICE*, News Bulletin of the International Glaciological Society, 1972, 16.)

21. For a description of the Harriman Expedition of 1899 and the resulting publication, see Chapter 2, note 8.

22. For a brief history of the boundary survey work along the Canada/Alaska border, see Appendix E.

23. Trimetrogon photography is a system of photographic mapping from the air in which one camera photographs the area below vertically and two cameras photograph obliquely (to the right and to the left) simultaneously at regular intervals.

24. The biographical information for Lawrence Martin is given in Chapter 3, note 22.

25. See Chapter 6 for Bill's mention of that trip.

26. A short biographical sketch of Wentworth and Ray can be found in Chapter 3, note 17. Bill mentions their 1931 trip to Prince William Sound in Chapter 6 and the publication resulting from that trip is given in Chapter 6, note 16.

27. A biographical sketch of François Matthes can be found in Chapter 6, note 23.

28. A short biographical sketch of Will Wright is given in Chapter 3, note 16.

29. The biographical information for H.F. Reid can be found in Chapter 3, note 33.

30. Austin Post (1922–) grew up in Chelan, Washington, and according to Bill Field, is "a man of great ability in various things. He is best known for his aerial photography of glaciers in various parts of Alaska, western Canada, and the western lower 48 every summer from 1960 to 1982. Prior to that, he served in the navy, worked for the U.S. Forest Service, worked on construction, and spent two years in the boat-building business. Even in those days, he was passionately interested in glaciers and changes in the glaciers. He began taking aerial photos of glaciers in Alaska in 1960 sponsored by the National Science Foundation and administered by Phil Church of the Department of Meteorology and Climatology at the University of Washington. Phil saw the importance of aerial coverage of glaciers and supported Austin's photo trips to Alaska and in the western United States and Canada in 1960, '61, '62 and '63. Austin joined the U.S. Geological Survey in 1963 and continued his aerial photography until his retirement. Austin's record of photography is unique in the sense that from 1960 to 1982 he generated annual aerial photography, both vertical and oblique, of an enormous number of glaciers over great stretches of country. Not only does Austin know glaciers and appreciate what features should be photographed, and how and to what detail, but also he is a good observer. This makes a big difference. He can take aerial pictures as well as interpret what he sees. That is perhaps the most valuable part of his contribution. It is nice for me to be able to record the appreciation of a colleague of this type, who has worked hard and consistently. One also should congratulate the Geological Survey for supporting a long-term project such as this air photography." (William O. Field, personal communication, 1990).

31. Mercanton, *Measurements of the Rhone Glacier 1874 to 1915* (1916).

FIGURE 269

CHAPTER 9

A BENCHMARK
IN GLACIOLOGY
1964–1980

Good Friday, 1964

The time between the end of the IGY and my retirement from the AGS in the late '60s was spent on two major projects—an earthquake investigation and compiling and editing a book. On March 27, 1964, Good Friday, the big earthquake in Prince William Sound occurred. The epicenter was between Columbia Glacier and Unakwik Inlet, where Meares Glacier is located (map 27). It shook the whole area (figs. 269–274), and very seriously damaged a lot of buildings in Anchorage. Valdez was very heavily hit. Most of the loss of life occurred not where the buildings were damaged, but along the coast where the huge waves that were sent up by the earthquake, presumably from submarine landslides in the fiords off the coast, came rolling into these fiords.[1] In Valdez, there were about thirty people on the dock watching a freighter that was just about to land and unload cargo

when the earthquake hit. The dock collapsed, and many of the people drowned. This is a tanker route now, so you have to remember what happened then can always happen again. Those waves were measured to have swept up to thirty meters on the side of the fiord at the Valdez Narrows (map 27).

I was a member of the Panel on Hydrology for the Alaska Earthquake.[2] We first had a tour of the damage in Anchorage. Then the air force took us on an aerial tour of Prince William Sound. We saw where the coast had submerged seven to ten meters. In particular, we were looking for whatever changes the glaciers might have undergone. We found very little to suggest that the glaciers had been affected,[3] other than an immediate calving process.[4] I think one reason the glaciers weren't affected very much was because it was early in the season, and the ground below and the water within the glaciers were still frozen. If the earthquake had happened

FIGURE 269.
Scene from Earthquake Park, Anchorage, Alaska. Aug 28, 1964. (F-64-K943)

FIGURE 270.
Another scene from Earthquake Park, Anchorage, Alaska. Aug 28, 1964. (F-64-K936)

FIGURE 271.
Some of the railroad cars along the waterfront in Seward which were damaged by a tsunami induced by the March earthquake. Sep 11, 1964. (F-64-K902)

FIGURE 272.
The remains of the ferry dock in Seward after the earthquake. Sep 11, 1964. (F-64-K896)

FIGURE 273.
Max Wells' former store on Alaska Avenue in Valdez after the earthquake. Sep 29, 1964. (F-64-K504)

FIGURE 274.
The Million Dollar Bridge after the earthquake. Note the collapse of the farthest right section of the bridge. July 25, 1965. (Photo by Austin Post, USGS, F654)

FIGURE 270

FIGURE 271

FIGURE 272

FIGURE 273

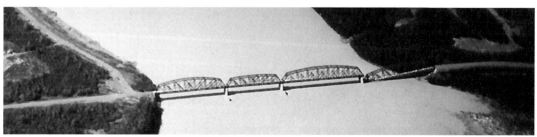

FIGURE 274

in September when there is a lot of water circulating through the lower part of the glaciers, it probably would have done much more damage. There were also changes in Cordova. There the land rose and the whole boat harbor was suddenly drained, but that was mainly a problem with where you could keep your boat.

There had been a big landslide on the Sherman Glacier outside of Cordova (map 27) during the earthquake, but it was covered with snow when it was photographed in April by the Arctic Institute [of North America], so we couldn't determine the extent of it at that time. We returned in September[5] when the lower part of the glacier is free of snow. It was a huge slide, much bigger than we'd ever anticipated from the earlier pictures (fig. 275). Debris a meter thick covered the surface of much of the lower end of the glacier and came within 400 meters of the terminus. We also saw the peak where the slide had originated, later named Shattered Peak by George Plafker, a geologist with the Alaska geology branch of the Geological Survey. In 1965[6] we started a systematic survey of the rate of flow of the ice, the extent of the slide, and the movement of the slide down-glacier. It kept moving toward the terminus, and by 1980 the slide reached the terminus. Several publications resulted from this study.[7]

Vigorous Advance or Catastrophic Recession?

Over the next twenty years, I made nine trips to Alaska in an attempt to keep track of the changes in the glaciers.[8] The glaciers in Alaska are very large and very dynamic, and their changes are so great that in just a few years one could miss a whole episode in the history of change of the glacier. In the Alps, one measures advance and retreat by a meter or so a year. In Alaska it can be up to tens of meters where there is a vigorous advance or hundreds of meters where there is a catastrophic recession. In '66, we were concerned with what the effects of the 1964 earthquake might have had on the glaciers. We made a whole tour of Glacier Bay and Prince William Sound. Except for the one

FIGURE 275A

FIGURE 275B

FIGURE 276A

FIGURE 276B

FIGURE 277

FIGURE 278

FIGURE 279

FIGURE 280

FIGURE 275.
Sherman Glacier before the slide (A), Aug 26, 1963 and after the slide (B), Jul 25, 1965. (Photos by Austin Post, USGS, K634-70 and F654-189)

FIGURE 276.
Harriman Glacier from a few meters SW of Sta. 2 looking WSW (A), Oct 1, 1931 (F-31-514) and from Sta. 2 looking WSW to the nearest point of the terminus (B), Sep 15, 1964 (M-64-P199). We measured a distance of 98 meters from the station to the ice front in 1964.

FIGURE 277.
At Johns Hopkins Sta. 5, July 24, 1971. Left to right: Jim Sanders, NPS boat skipper; Bob Howe, Glacier Bay National Park Superintendent; Marion Millett, AGS; Austin Post, USGS. (F-71-K305)

FIGURE 278.
At Muir Sta. 32, July 26, 1974. Bob Howe (L), Dick Ragle, and Dave Bohn (R). (F-74-K188)

FIGURE 279.
In Johns Hopkins Inlet in 1976. John Field (L), Bob Howe (retired), Rick Caulfield, and Superintendent Tom Ritter (R). (F-76-K348)

FIGURE 280.
Party on the Nunatak in Reid Inlet in honor of the 50th Anniversary of my first visit to Glacier Bay. Left to right: Tom Ritter, Jim Sanders, Ken Youmans (standing), Bruce Paige (kneeling), John Field, Bob Howe, Gene Chaffin, Rick Caulfield, and Chuck Janda. Aug 6, 1976. (F-76-K418)

FIGURE 281.
*Jim Luthy and Gene Chaffin
from the* Nunatak *on the
dock at Bartlett Cove,
Glacier Bay, 1982.
(F-82-K144)*

FIGURE 282.
*Dave Bohn and I in Hugh
Miller Inlet, Sep, 1966.
(Photographer unknown;
from the WOF Collection.)*

FIGURE 283.
*Dick Goldthwait at Sta. 1A
in Johns Hopkins Inlet. He
was professor of geology at
Ohio State University in
Columbus, Ohio. Aug 21,
1958. (F-58-K88)*

FIGURE 281

FIGURE 282

FIGURE 283

noticeable effect at Harriman Glacier (fig. 276), I didn't detect anything special. In '67 we went back to the Coast Mountains and Glacier Bay. In '68, we redid Glacier Bay, looking for the effects of the 1958 earthquake down there. And in Prince William Sound, we continued to look for effects of the '64 earthquake. We returned for a brief trip in '71 on the way to a meeting in Moscow. I found that for $200 extra I could go to Moscow by way of Alaska instead of straight from New York, so en route I spent a week in Glacier Bay and a week in the Anchorage area. In '74 I returned to Prince William Sound and Glacier Bay. I did not go to Columbia Glacier because it was being studied by the Geological Survey. The glacier was still as I had seen it, with minor changes, during my first visit in 1931, which was very much as Harriman had seen it in 1899.[9] My last major field trip to Alaska was in '76, the fiftieth anniversary of my first trip to Glacier Bay (figs. 277–284) and I visited both Glacier Bay and Prince William Sound.

Mountain Glaciers of the Northern Hemisphere

In the middle '60s, Natick Laboratories of the U.S. Army Quartermaster Corps, which had sponsored the *Geographic Study of Mountain Glaciation in the Northern Hemisphere*, asked me, as head of the Department of Exploration and Field Research, if we would consider doing a revision of the manuscript. It turned out that very little of the old parts could be used, plus there was a great amount of new material that was added during the IGY, so they supported not a revision, but a complete rewrite. This took years to do, and rather swamped us. The last chapter was finished in 1972 and the three-volume set was published in 1975 by the U.S. Army Cold Regions Research and Engineering Laboratory (CRREL)[10] at the urging of Dr. Malcolm Mellor, a scientist at CRREL. I remember the charge was twenty dollars for the two volumes of text and one of maps. It was titled *Mountain Glaciers of the Northern Hemisphere*,[11] and I was rather pleased to be called "the editor." This was a major effort. It took an enormous amount of time. Some people have called the work a benchmark and, in a sense, I guess it is. It is difficult for me to gauge and say what value it has, but I think I've received enough remarks to feel that quite a lot of people consider it a valuable reference. The purpose of the publication was to provide future researchers access to what has gone on before without having to go back to all the original data. We were not concerned with the ice sheets, which are subjects in themselves. If we had gotten into that, we'd still be trying to write and assemble the manuscript.

The 1975 publication has not been renewed, but now one can study many of those same features as well as the distribution of glaciers by satellite imagery. The resolution of these images is getting better and better and, while it doesn't take the place of detailed ground work, one can see a great amount of detail in these images (fig. 285).

Retirement from AGS

By the mid-'60s, I was supposed to be working almost full time as editor of *Mountain Glaciers of the Northern Hemisphere*, as well as trying to do other projects. I had been telling the people at the AGS that they must get somebody to take my place. The person would start as my assistant, and then I would gradually relinquish all responsibility as head of the Deparptment of Exploration and Field Research to the person so that I could finish editing *Mountain Glaciers of the Northern Hemisphere*. In 1964, I wrote a memorandum saying we would not take on any new projects. We couldn't do the

FIGURE 284

FIGURE 284.
My son John, shown here in 1976, joined me on many trips to Glacier Bay. (Photographer unknown; from the WOF Collection.)

FIGURE 285.
LANDSAT imagery of Glacier Bay, Alaska, Sep 12, 1973. (From the WOF Collection)

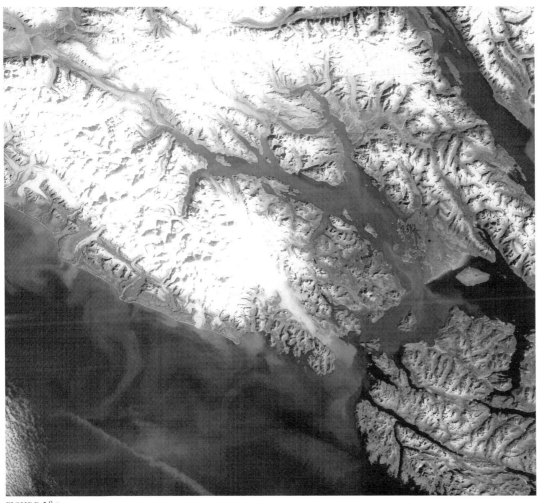

FIGURE 285

FIGURE 286.
Our house in Great Barrington, Massachusetts, June 1993. (Photo by C S. Brown.)

FIGURE 287.
My office, the remodeled carriage house, June 1993. (Photo by C S. Brown)

FIGURE 288.
Bill at work in his office, June 1993. (Photo by C S. Brown.)

FIGURE 286

FIGURE 287

FIGURE 288

ones we had, and I had nobody to turn the work over to on a gradual basis before I retired. I finally retired as a paid member of the staff of the AGS in 1968, several years earlier than planned in order to finish *Mountain Glaciers of the Northern Hemisphere*. I did this deliberately because I thought it would force some action. It would wake the Society up to the fact that our department just couldn't go on anymore. My retirement, however, forced no action and thus ended the active participation of a research operation within the American Geographical Society.

By the 1970s the old AGS building at Broadway and 156th Street that the Society had occupied since 1912 was in an awkward part of the city. The city had not grown northward as it had been expected to do in the early part of the century, when Columbia University was at 115th Street. The AGS building was originally the house of the scientist James Audubon, and was converted into a center composed of the Academy of Arts and Letters, the Numismatic Society, the Hispanic Society, the Museum of the American Indian, and the AGS. The AGS sold the old building and now has an office on Fifth Avenue. Prior to selling the building, the AGS library with its invaluable collection of books and maps was transferred to the Golda Meir Library at the University of Wisconsin, Milwaukee. The AGS material is still there and being managed successfully and with enthusiasm.

The Department of Exploration and Field Research, including our maps and reading material, was always a separate unit within the AGS. When the AGS library moved to Milwaukee and the AGS offices to a new building, no one was interested in the department's material. Instead of allowing the material to be thrown out, I decided to take it. In April of 1980, I had four weeks in which to clear everything out. During two weeks of that there was a New York City subway strike and during the other two weeks I had an infected tooth. I had a moving van back up to the front door of the office, I loaded it up myself, and had the material brought to my new home in the Berkshires.

Return to the Berkshires

I left the Berkshires in the late '20s, and moved back after more than fifty years. In the late '70s, Mary and I started to look for land in this general area. We wanted to be in the Berkshires, but far enough away from where I grew up so I didn't feel the connection to the gilded life of my youth. My sister lives about ten miles away, and we're about thirteen or fourteen miles from Tanglewood[12] and Lenox, yet we are out of the great traffic jams of midsummer in the Lenox area. Whenever there's a performance at Tanglewood involving ten to twelve thousand people the roads tend to get clogged up, and Stockbridge is especially hard to get through. I've often said it's easier to go through Times Square on an ordinary day than through Stockbridge in midsummer. Until a couple of years ago, we kept an apartment in New York for convenience and pleasure, as Mary enjoyed going to the theater and concerts there every once in a while.

The old house we bought and remodeled included on the property what had been a carriage house in the old days. The upper part of the building, once a hayloft, we had remodeled into a small apartment. Downstairs, where the carriages used to be housed, became my office (figs. 286–288). There were three horses there when we bought the property in 1977 and after the horses were removed, we made another room out of horse stalls. This is where I keep all my negatives and slides. So this carriage house is where my collection from over the last sixty years is housed. And when I pass on to less activity, or no activity, I want to be sure that this material is in a place where it will be preserved as well as accessible to others, and I hope added to as time goes on.[13]

Notes

1. In some cases the waves originate at the head of the fiords and wash out (Austin Post, personal communication, 1988).

2. Members of the Panel on Hydrology for the Alaska Earthquake were Richard Goldthwait, Chairman, Ohio State University; William O. Field, American Geographical Society; Troy L. Péwé, Arizona State University; Robert C. Vorhis, U.S. Geological Survey; and Roger Waller, U.S. Geological Survey.

3. Field, W.O., "The Effect of Previous Earthquakes on Glaciers" (1968).

4. The calving process is the process by which masses of ice break away from the terminus of a glacier or an ice sheet that ends in a body of water. The large pieces of ice that have broken off are called icebergs.

5. See Appendix C for a summary of the trip.

6. See Appendix C for a summary of the trip.

7. See Clayton, W.O. Field, and Tuthill, "Recent Fluctuations of the Sherman and Sheridan Glaciers, South-central Alaska" (1966); Field, W.O., "Avalanches caused by the Alaska Earthquake of March 1964" (1965); Marangunic and Bull, "The Earthquake-induced Slide of the Sherman Glacier, S. Alaska, and its Glaciological Effects" (1966); Marangunic and Bull, "The Landslide on the Sherman Glacier" (1968); Plafker, "Source Areas of the Shattered Peak and Pyramid Peak Landslides at Sherman Glacier" (1968); Shreve, "Sherman Landslide, Alaska" (1966); Tuthill, "Sherman Glacier, Paleoecologic Laboratory"

(1966); and Tuthill, W.O. Field, and Clayton, "Postearthquake Studies at Sherman and Sheridan Glaciers" (1968).

8. See Appendix C for a summary of these trips.

9. For previous terminus positions of Columbia Glacier, refer to map 29.

10. U.S. Cold Regions Research and Engineering Laboratory, 72 Lyme Road, Hanover, New Hampshire 03755.

11. Field, W.O., *Mountain Glaciers of the Northern Hemisphere* (1975).

12. Tanglewood, an estate in Lenox, Massachusetts, is the summer home of the Boston Symphony Orchestra. It was founded as the annual summer Berkshire Symphonic Festival (now the Berkshire Festival) in 1934 by Serge Koussevitzky (1874–1951), conductor of the Boston Symphony Orchestra. The festival consists of outdoor performances by the orchestra. In 1940 Koussevitzky founded the Berkshire Music Center, a summer music school also at Tanglewood.

13. The William O. Field Collection was packed and shipped to the archives in the Alaska and Polar Regions Department at the Rasmuson Library, University of Alaska Fairbanks in November, 1993. So readers may better understand the magnitude of the collection as well as the amount of work that Bill put into it, following is a brief description of the collection:

    • 775 books, primarily covering the Arctic and Antarctic, early mountaineering with some from Europe written in the early 1800s, Alaska (general topics), glaciology and other earth sciences.
    • Clippings dating back to 1947 relating to projects of the Department of Exploration and Field Research or the polar regions.
    • 17 boxes of magazines from the 16 subscriptions Bill held.
    • 4 file cabinets of reprints by other authors arranged by the topics:
        glaciology–general
        glaciology–North America
        glaciology–outside North America
        Polar Alaska–nonglacier
        climatology
        geology–glacial
        geology–Quaternary hydrology
        general science and geography
        expeditions and mountaineering
    • 72 large photo albums of the best pictures of glacier termini taken by Bill beginning in 1926 and ending with his last

trip in 1987, organized both by year and by glacier.

• Over 1,500 photographs of glaciers in Alaska taken by other people with half being terrestrial photos taken as far back as the 1880s and the remaining being aerial photos from the 1929 U.S. Navy flights, the 1941–1942 U.S. Army Air Corps trimetrogon flights, the 1948 U.S. Navy flights and aerial photos taken by Austin Post of the U.S. Geological Survey from 1963–1980s.

• 51 book boxes of negatives and slides numbering in the thousands, glass slides, and movie film taken on Bill's field trips.

• Field diaries, field notes, histories of station occupations dating back to the boundary survey work of 1893; unpublished manuscripts and correspondence.

• 16 full-sized map drawers containing old Alaska maps dating from the 1890s, Bill's survey and sketch maps and/or U.S. Geological Survey topo maps with his surveys drawn on them, Alaska topo maps at 1:250,000 and 1:63,360 scale, maps from the Canadian Rockies composed of ones he collected, ones he used in the field, and ones he used back at the AGS to work up his survey data.

The collection is slowly being processed and parts of it, especially Bill's glacier photo albums and the photographs of glaciers taken by others, should be available for public use in the near future.

FIGURE 289

CHAPTER 10

THE RACE WITH TIME
1980–1994

As I reflect back on my life and career, I realize how lucky I have been. When people seriously began to study glaciers in the 1940s, I was already involved in glacier monitoring as a hobby, beginning of course in 1926, and then I moved right into a job as a glaciologist with the AGS. I guess you could say I was in the right place at the right time. More important, I feel fortunate that I knew many of the early people—Martin, Reid, Cooper, the brothers F.E. and C.W. Wright, Mertie, Capps, Moffit and Merriam to name a few. I think my main contribution to glaciology has been bridging the gap between the early work of these people in Alaska beginning in the 1880s until the end of the boundary survey work in 1914, and the modern-day research begun after World War II. I think one of the more interesting items I can contribute is a description of the people who took part in those early trips, what they did, and how they did them. For instance, they *rowed* boats to the glaciers. John Muir describes crossing Muir Inlet in an hour or so, rowing through the ice after a long day's work. It was a tough operation. Gilbert of the Harriman Expedition rowed, and

we even used oars on most of our '26 trip. Jim Huscroft in Lituya Bay had a little outboard that we used, and old Burroughs in Taku Inlet towed us in Percy Pond's canoe here and there with his little outboard. By '31 Andy Taylor and I had an outboard at Columbia Glacier and that made a big difference. So, it has been an extraordinary change, going from rowing our boat in '26 to the use of helicopters today, to the use of satellites in the near future for glacier monitoring. I guess I have to agree that I have spanned the whole spectrum of glacier monitoring.

In addition to his enthusiasm and encouragement, one of the treasures that Bill brought to many of us was his personal knowledge of the pioneers in glaciological research.

Carl Benson

FIGURE 289.
Friends and colleagues still stopped in to see me in Great Barrington. This is Carl Benson from the Geophysical Institute at the University of Alaska Fairbanks with me in 1984. (Photo by Ruth Benson.)

Over my years of glacier observations, in general it has been a fight to convince people of the importance of long-term systematic glacier monitoring and to obtain funding for glacier studies. The mass balance of glaciers is taken for granted and the thought that a glacier might disappear or double in size is not considered very likely. But small glaciers react almost immediately to changes in the amount of snowfall or the rate of melting. Information from high-altitude meteorological stations near a glacier would help greatly in the study of the glacier/climate relationship, but there are very few of them due to the high cost of establishing and maintaining such stations. In general I feel I've had good cooperation with regards to my glacier monitoring work from the folks in Glacier Bay National Park. The park is squeezed for money just like all the other national parks. In this country, long-term planning is very difficult to achieve. It tends to be within the four-year term of a president or a congressman or a legislator, and monies tend to be spent on seemingly more urgent short-term problems. Thus developing a long-term policy, such as in Switzerland or Austria, is very difficult.

In assessing my contribution, I think my work has been well received by our friends in the fraternity but I don't think I have made any great impact on the world. I still receive a lot of requests for information but that is what happens if you live a long time. Henry Fonda once said that if you live long enough, you become famous without having to do much of anything. If you can remember events from long ago, you don't have to have been a very profound observer to say something useful!

Even before 1930 I was fascinated with the question of what the glaciers were doing. Martin last recorded the state of the glaciers in 1909, and his listing of what stations to occupy in the future interested me. I also discovered that the behavior of the glaciers in Prince William Sound was so different from that of the glaciers in Glacier Bay. It was an interesting problem which a person with limited technical training like myself could help answer. I enjoyed Alaska, I enjoyed the work, and I felt that sooner or later my effort to continue to record the positions of the glacier termini would be a contribution, if for no other reason than that they were not being recorded by anyone else. The idea of global warming hadn't been thought up yet, but we knew something was happening and any records would become very interesting and valuable. An interesting fact about global warming is that there would be a warmer North Pacific Ocean and thus warmer air masses, which hold

more moisture. That would result in more snow in the higher elevations of the coastal mountains and thus, initially, glacier advances even with a warming trend. Before you can solve a problem, you have to know what's happened before, and a photographic record tells so much. Many times I thought how great it would have been if we had had this sort of record from two hundred years ago. Besides providing a record that would prove useful in the future, I feel my work has been like counting pebbles—it refines the measurements made by airplanes and satellites and somebody has to do it—somebody still has to count the pebbles.

The biggest regret I have is that I feel I could have worked much more effectively. We worked hard out in the field, but the material was never adequately put together when we came back to the office. I'm still trying to do that because I remember a lot that is not in the records that I want to write down, but it is a matter of time, time, time. I become involved in other things, like answering inquiries that come in. That takes me away from my work, but I do not feel that it is wasted time. I always took the position in working for a nonprofit organization that it was our job to try to answer all inquiries. There are the people who are beginning to study a glacier or a glacierized area and want to know what is the history of observations. If I can supply some of that information, it's helpful. I frequently think of the statement by François Matthes, "Nothing makes me happier than when I can share the results of my scientific studies with laymen." That's the motivating factor. That's why I try to answer the letters. Of course, there is a balance that has to be hit between all these things that I need to do. I wake up at night and worry about this. Because my time is limited, how should I mix things adequately? There are the things that I know I must do versus the things that I feel I should do. So this is a problem, a serious problem. It's a race with time. How many more trips to Alaska can I make? I'm not sure. I don't think there will be very many. But I'd like to go back. I'd like to see what the glaciers are doing.

The snows that are older than history,
the woods where the weird shadows slant;
The stillness, the moonlight, the mystery,
I've bade'em good-by—but I can't.

Robert Service

IN MEMORIAM
WILLIAM OSGOOD FIELD
January 4, 1904–June 16, 1994

In 1926, the face of Muir Glacier was just north of what is today known as Goose Cove and alders were beginning to creep around John Muir's cabin at Muir Point. It was in this year that a young adventurer named Bill Field made his first visit to Glacier Bay and fell in love with this raw, dynamic landscape. Later, reflecting on his visit, reading the works of Muir and other early visitors, and personally discussing the Bay with William S. Cooper and Harry Fielding Reid, Bill soon realized the immensity of Glacier Bay's potential as an exemplar of glacial processes, and therefore as a window into the Ice Age past of North America.

Bill Field's relationship to Glacier Bay became a testament to the value of clarity and consistency of purpose. His genius, and enduring gift to the future, was to become convinced that Glacier Bay's utility as a natural laboratory fundamentally depended on knowledge of the extent of glacial ice over time—and then to resolutely act upon this conviction throughout a long lifetime. For almost seventy years he was the prime repository of information on glacial extent along the Bay, making many trips from his East Coast home at personal expense to gather the information first-hand. When, in later years, age prevented his return to the Bay to personally re-occupy his photo stations, he relied on others to do so and turned his attention to archiving his voluminous photo collection, which he donated shortly before his death to the University of Alaska.

Bill was a fundamentally modest man. He actively avoided the limelight and seldom joined the company of leading scientists of the day. To give a lecture or deliver a paper was an ordeal to which he seldom succumbed. Such works as he committed to paper were carefully restricted to what he had seen and could vouch for; he left theory to those he considered better suited for its pursuit. It turned out to be fortunate that Bill was not a highly trained scientist. Otherwise he would have become too captivated by theoretical questions to devote the effort he did in pursuit of so simple a goal —recording the extent of glaciation through time.

As much as I am grateful to Bill for his professional contribution, my major joy in his recollection comes from reflections of Bill Field the man. To be near Bill was to feel the power and beauty of his sense of place, acted upon in the context of profound respect. He was not one to wear his personal feelings on his shirtsleeve, or to bare his soul in an evening's conversation. You had to read the depth of his soul in his clear, gentle eyes, and in the courtesy and dignity with which he carried out each personal transaction. Seated with Bill in the *Nunatak*'s galley, or looking out with him across the landscape from one of his beloved survey stations, you knew instinctively that you were in the presence of a great gentleman. His passing leaves a hole no one of us can fill. But as I think Bill would say with a small smile, "The glaciers are still there."

Greg Streveler, Glacier Bay, August 1994
for *Friends of Glacier Bay Newsletter*

COMPARATIVE GLACIER PHOTOGRAPHS

This section displays some of the most spectacular expamples of comparative glacier photography in Bill's collection. These photographs not only are beautiful but also represent one of his major contributions to glaciology in North America. Besides highlighting Bill's own photographs beginning with his first trip in 1926, this section includes some photographs in William O. Field collection taken by other people visiting Alaska, most around the turn of the century. Bill's glacier monitoring program investigated glaciers as living processes whose change over time holds secrets to fluctuations in climate and ecosystems both large and small. These photographs may be enjoyed both for their beauty and as part of the enduring scientific legacy of a pioneering glaciologist.

1A, 1B, 1C
VIEWS NORTHWESTERLY FROM REID'S STA. V
Reid's Sta. V (C. W. Wright's Sta. 11 in 1931) is at an elevation of 3000 ft
on the northern shoulder of Mt. Wright. The 1850 ft ridge which was
Reid's Sta. G is near the center of the photo. This ridge, which was a nuna-
tak in the 1880s, was also popularly referred to at that time as The Rat.

1A. Aug 19, 1892, H.F. Reid (No. 381). Morse Glacier center left, Muir terminus lower right.

1B. 1907, E. Martin (IBS) (No. EM-59-07). Morse Glacier center left, Cushing Plateau center, Muir Glacier right.

1C. 1931, C. W. Wright. Morse Glacier (center) and Plateau Glacier (upper right).

2A. 1929, U.S. Navy, National Archives (No. 80-CF-781381-13).

2B. July 25, 1968, W.O. Field (No. F-68-K279).

3A, 3B
NORRIS GLACIER FROM TAKU STA. 4

3A. Aug 1926, W.O. Field (No. F-26-18).

3B. Sep 7, 1950, W.O. Field (No. F-50-R707, R708).

4A, 4B, 4C
THREE GLACIERS VIEWED FROM STA. WELLS
Lamplugh (top left), Johns Hopkins (top right) and
Grand Pacific (bottom, first photo only) Glaciers from
Sta. WELLS, also known as Sta. 194.

4A. 1894, A.J. Brabazon (A,B,C) (No. 44).

4B. 1907, Netland (IBC) (No. N-82-07).

4C. 1972, Bob Howe (NPS).

RENDU GLACIER FROM AGS STA. 4

Station first occupied by H.F. Reid in 1892 (Reid's Sta. A).

Also known as USGS Sta. HUMP.

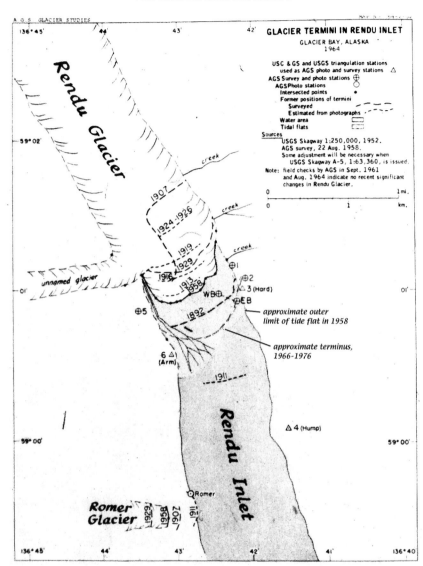

5A. Map of glacier terminus positions in Rendu Inlet
(map prepared by W.O. Field at the AGS).

5B. 1892, H.F. Reid.

5C. 1907. Note on map big retreat since 1892. E. Martin
(IBS No. E.M.-66-07) (E. Martin No. 9218).

5D. Sep 1, 1911, Lawrence Martin. Major advance and farthest recorded extent of Rendu Glacier (No. 27-1-1911).

5E. Aug 27, 1916, W.S. Cooper. Recession of 1.5 miles since 1911 (No. 20-4773).

5F. Aug 12, 1950, W.O. Field. Slight advance since 1916 (No. F-50-B65).

5G. 1961, AGS photo by M. Millett. Much thinning and slight retreat since 1950 (No. M-61-P136).

5H. 1966, W.O. Field. Vigorous advance in five years (No. F-66-P41).

5I. Aug 6, 1997, C S. Brown. Much thinning and slight retreat. Notice the trim line on both the right and left sides of the glacier terminus (No. CSB11-AK97-8, 10).

6A, 6B
GRAND PACIFIC GLACIER
From AGS Sta. 6 (called Tarr Sta. C in 1926).
Arrow points to the same glacier.

6A. Aug 28, 1926, W.O. Field (No. F-26-231).

6B. Aug 14, 1950, W.O. Field (Nos. F-50-495, 496, 497).

7A, 7B
HUGH MILLER GLACIER FROM STA. A

7A. 1926, W.O. Field (Nos. F-26-180, 181).

7B. 1941, W.O. Field (Nos. F-41-462, 463).

8A, 8B
GRAND PACIFIC GLACIER
From Sta. WELLS on the east side of the entrance to Tarr Inlet.

8A. 1894, A. J. Brabazon (Nos. 44–47).

8B. 1907, Netland, No. N-82-07. Grand Pacific Glacier retreated more than 7.5 miles from 1894 to 1907.

9A, 9B, 9C
Muir Glacier
From the summit of the Nunatak
(AGS Sta. 1) looking Northwest. Arrows
point to same rock for reference.

9A. 1936, John Reed, USGS (No. 234).

9B. 1941, D.B. Lawrence (No. 41-IV-7).

9C. Aug 9, 1950, W.O. Field
(No. F-50-R281).

10A, 10B, 10C
MUIR GLACIER FROM STA. 3.
View NNW. This was called Sta. A by
W.O. Field in 1926 but was changed to
Sta. 3 by C.W. Wright in 1931.

10A. 1929, W.S. Cooper.

10B. 1941, W.O. Field (No. F-41-38).

10C. 1950, W.O. Field (No. F-50-R290).

HISTORICAL PHOTOGRAPHS

After Bill's 1931 trip to Alaska he realized there was no one inclusive collection of glacier-related photographs. He decided to assemble as many photographs as possible taken by people on earlier expeditions; this section contains a sample of these early photos. They illustrate how much travel to Alaska there actually was around the turn of the century. Muir Glacier in Glacier Bay was a popular tourist attraction, with the first tourist party visiting the glacier in 1883. These photographs are both enjoyable to look at as well as historically significant and illustrate the breadth of interest and knowledge of William O. Field.

1. Reid, Johns Hopkins, and Grand Pacific Glaciers (clockwise from bottom) in 1894 from Sta. KING (also known as Sta. SLIP). Photo by A.J. Brabazon, Alaska Boundary Commission (No. 77).

2. Aerial View of Taku Glacier taken by the U.S. Navy in 1929 (U.S. Navy #GS-284).

3. Tourists at Muir Glacier, about 1895. Photo by Veazie Wilson (No. 131).

4. Tourists at a crevasse on Muir Glacier, about 1896. Photo by Winter & Pond of Juneau (No. 170 or 179).

5. The first tourist party to go ashore at Muir Glacier, July 13, 1883. A landing was made on the west side of the inlet. One of the passengers was Eliza R. Scidmore, who later gave this picture to Lawrence Martin. Presumably it was taken by the ship's photographer.

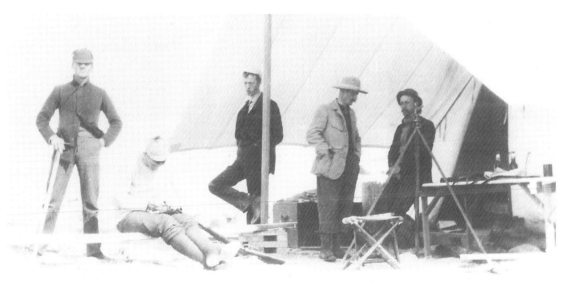

6. Reid's party in camp at Muir Glacier, 1890. Possible identification is (left to right) R.L. Casement, H. McBride, C.A. Adams, H.F. Reid, and H.P. Cushing. Photo probably taken by J.F. Morse (C30).

7. Two members of Reid's party, possibly H. McBride (left) and H.P. Cushing (right) at markers on a glacier (probably Muir) for measuring surface flow, 1890. Photo probably taken by J.F. Morse (C30).

8. Norris Glacier photographed by the commercial photographer Partridge. William Williams bought this postcard in Sitka in 1888, thus it shows the glacier in a pre-1888 position.

9. International Boundary Survey party of 1907 under Fremont Morse at Sta. WELLS at an elevation of 4105 feet on the east side of the entrance to Tarr Inlet. This site was first established by A.J. Brabazon of the Canadian contingent of the Alaska Boundary Commission in 1894. Photo by Netland (No. N-80-07).

10. A Siwash camp on the beach near Camp Muir. Photo by H.F. Reid, June 23, 1892 (No. 271).

11. International Boundary Survey party
of 1910 at the head of Carroll and
Tsirku Glaciers en route from Tsirku
Valley to Boundary Point No. 157.
Photo by O.M. Leland.

12. G.K. Gilbert's party from the
Harriman Alaska Expedition of 1899 in
Glacier Bay. E.H. Harriman photo,
June 1899 (No. 56).

13. A paddle-wheel steamer at Muir Glacier in the 1880s. Charles Hallock, who visited Muir Glacier in about 1885, wrote of the difficulty a paddle-wheel steamer had with ice: "the wheels were afterward badly smashed in making her way out of the bay into open water. A paddle-wheel steamer is unfit for such navigation." This postcard, a photograph by Partridge, was bought by William Williams in Sitka in 1888 (No. 7768c or g). The steamer is probably the *Ancon*, remembered at Ancon Rock just off Point Gustavus where it sank in 1889.

14. The U.S. Coast and Geodetic Survey ship *Cosmos* in front of Lituya Glacier. Photo by W.O. Field, Sep 11, 1926 (No.26-298).

15. S.S. *Queen* at Muir Glacier, about 1896 (Winter & Pond No. 272).

16. Muir Glacier and Muir's cabin (center) from east side of Muir Inlet, July 1896. Steamer appears to be the S.S. *Topeka* (Winter & Pond No. 272).

17. An unknown steamship in front of Columbia Glacier, Alaska (undated).

18. The steamer *George W. Elder* in Glacier Bay, June 1899, carrying the Harriman Alaska Expedition (E.H. Harriman photo No. 31).

19. The S.S. *Idaho* under the command of Captain Carroll at Muir Glacier, July 13, 1883. This was the first visit to Muir Glacier since that of John Muir and S. Hall Young in 1880 (no number).

20. The steamer *Spokane* at Taku Glacier, 1907. Photo by Case.

21. Harvard Glacier, 1905. Photo by
Paige, USGS (No. 693).

22. Harriman Glacier from the boat. Photo by the Harriman Alaska Expedition of 1899 (Curtis photo No. 294).

23. College Fiord. Photo possibly by Sargent of the USGS, 1916.

MAPS

This section includes all the maps referenced in the text. In addition to general area maps, many special maps from Bill's collection have been incorporated into his story and are included here. Almost half of these illustrations were drawn by Bill himself, either in the field or based on field notebooks. In addition to illustrating the routes of Bill's journeys and the scope of his work, historical maps are included that illustrate the landscapes encountered by his colleagues and predecessors. These provide critical documentation for Bill's studies of glacier change. The oldest map dates back to 1786; four are from the 1890s, and four are from the early 1900s. Like his photographs, these maps are part of the enduring scientific legacy of William O. Field.

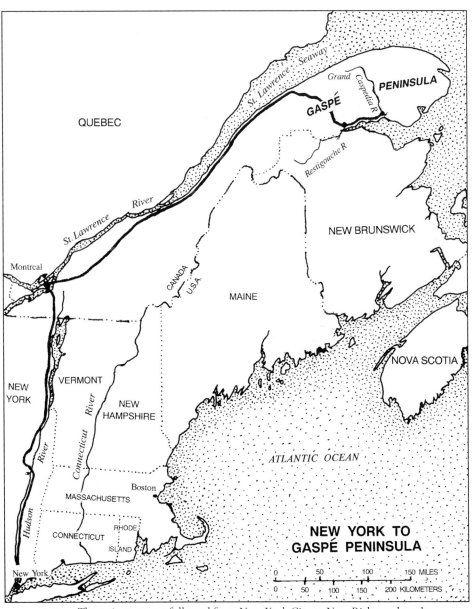

MAP I. The train route we followed from New York City to New Richmond on the Gaspé Peninsula of Quebec. © D. Molenaar 1997.

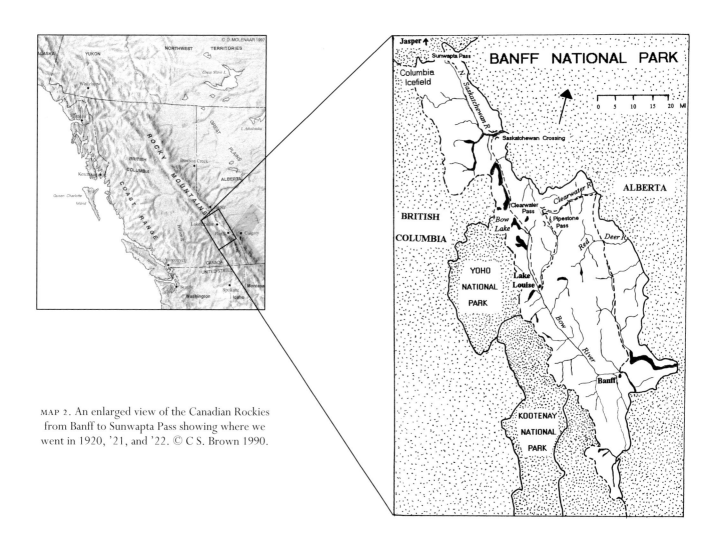

MAP 2. An enlarged view of the Canadian Rockies from Banff to Sunwapta Pass showing where we went in 1920, '21, and '22. © C S. Brown 1990.

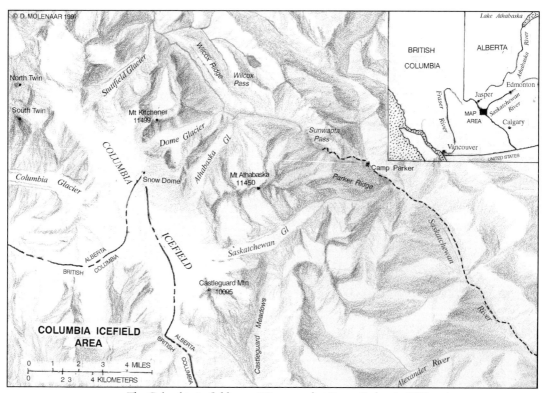

MAP 3. The Columbia Icefield area. We camped at Camp Parker in 1922 and Castleguard Meadows in 1924. © D. Molenaar 1997.

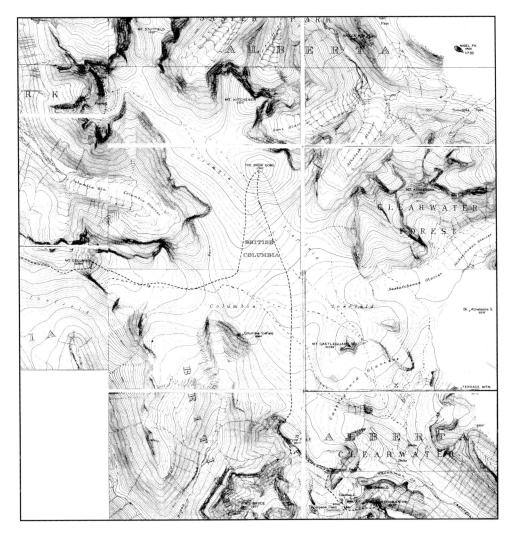

MAP 4. This is the map we used in 1924, with our climbing routes sketched in light dashed lines (the heavy dashed lines were published on the original map).

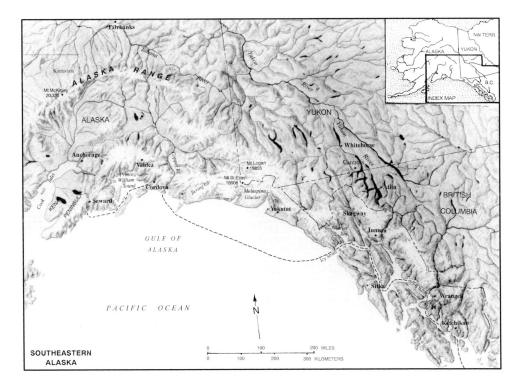

MAP 5. We took a boat up the coast from Seattle to Seward, stopping at the major ports, and then we boarded a train for the trip from Seward to Fairbanks.
© D. Molenaar 1990.

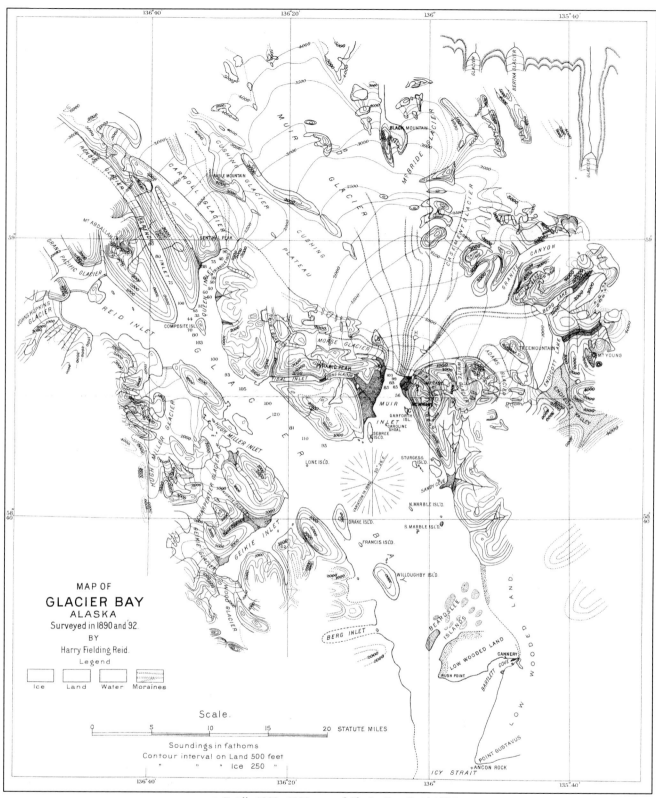

MAP 6. Reid's 1890–1892 map of Glacier Bay (Reid 1896).

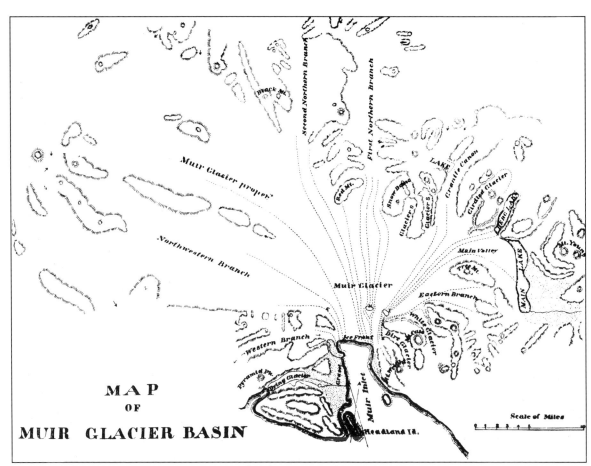

MAP 7. H.P. Cushing's 1891 map of Muir Glacier Basin.

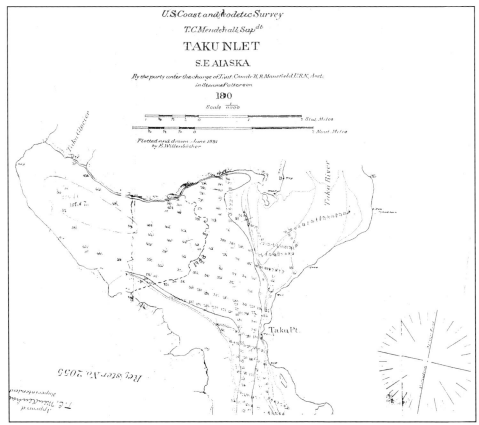

MAP 8. The 1890 USC&GS map of the terminus of Taku Glacier and the
bathymetry in upper Taku Inlet

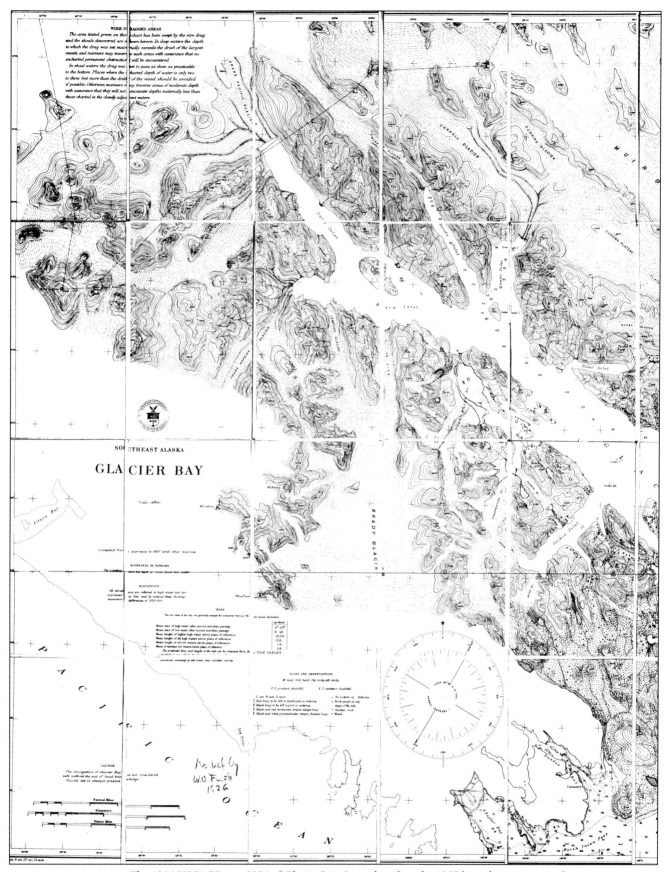

MAP 9. The 1916 USC&GS map 8306 of Glacier Bay. It was based on the 1907 boundary survey work.

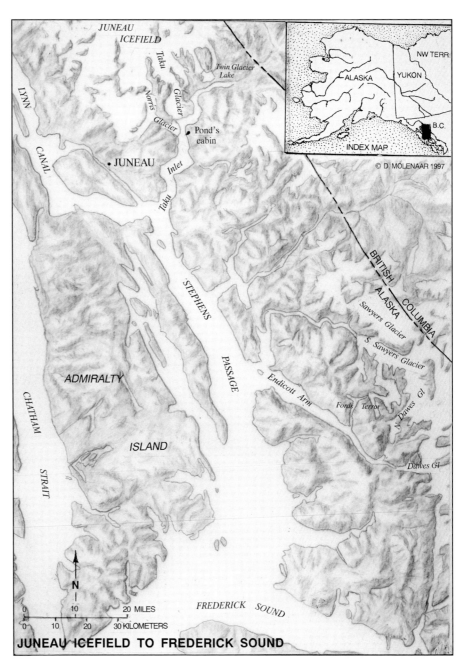

Map on right contains labels:
JUNEAU ICEFIELD
Twin Glacier Lake
Taku Glacier
Norris Glacier
Pond's cabin
LYNN CANAL
• JUNEAU
Taku Inlet
Taku
STEPHENS PASSAGE
ADMIRALTY ISLAND
CHATHAM STRAIT
Sawyers Glacier
S. Sawyers Glacier
Endicott Arm
Fords Terror
N. Dawes Gl.
Dawes Gl
BRITISH COLUMBIA
ALASKA
N
0   10   20 MILES
0   10   20   30 KILOMETERS
FREDERICK SOUND
JUNEAU ICEFIELD TO FREDERICK SOUND

Index map:
NW TERR.
ALASKA
YUKON
B.C.
INDEX MAP
© D. MOLENAAR 1997

MAP 10. Juneau and the surrounding area, including the head of Taku Inlet, the fiords south of Juneau and the Juneau Icefield northeast of Juneau. © D. Molenaar 1997.

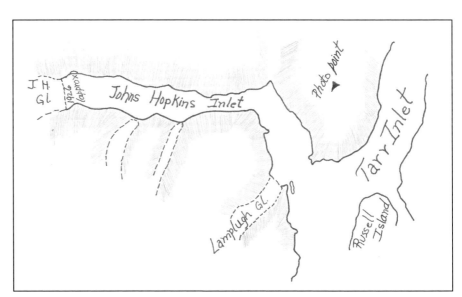

Labels on map:
J.H. Gl.
(1929&30)
Johns Hopkins Inlet
Photo point
Tarr Inlet
Lamplugh Gl.
Russell Island

MAP 11. Johns Hopkins Inlet showing the ridge in the mouth of the inlet from where I took my 1926 photos of Johns Hopkins Glacier.

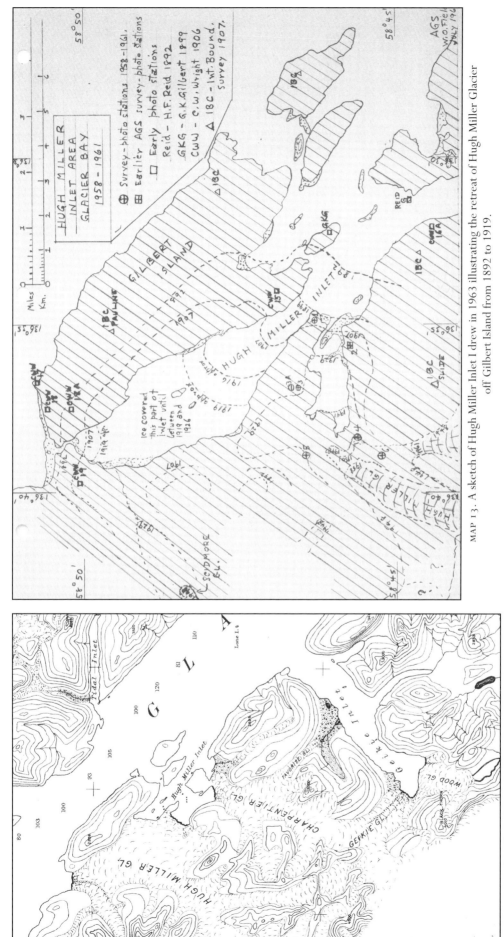

MAP 13. A sketch of Hugh Miller Inlet I drew in 1963 illustrating the retreat of Hugh Miller Glacier off Gilbert Island from 1892 to 1919.

MAP 12. Hugh Miller Glacier as it retreated off of what became Gilbert Island. This is a section of the 1899 USC&GS map 3095 of Glacier Bay compiled from surveys by the USC&GS, H. F. Reid, and the Alaska Boundary Commission between 1884 and 1896.

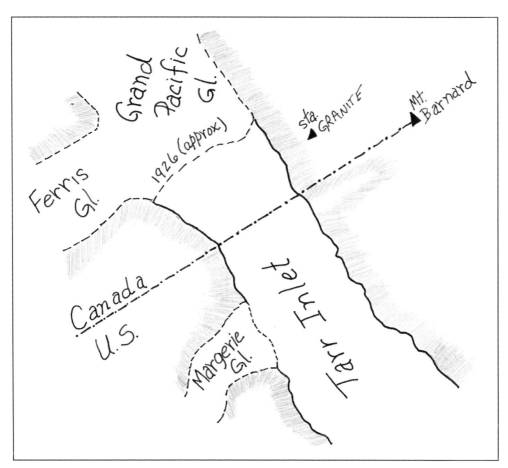

MAP 14. The head of Tarr Inlet in 1926. Grand Pacific Glacier was well back of the U.S.-Canada border then.

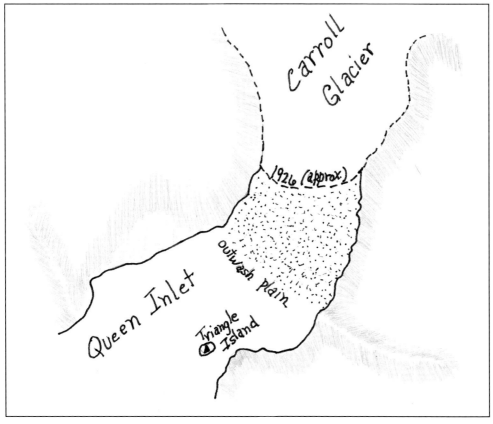

MAP 15. Queen Inlet and Carroll Glacier in 1926.

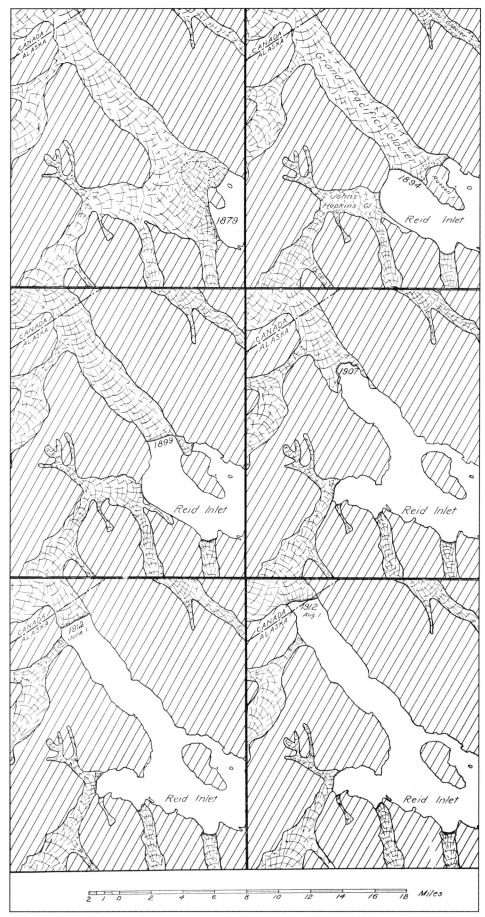

MAP 16. Grand Pacific Glacier and the opening of Reid and Tarr Inlets between 1879 and 1912 (from Martin 1913).

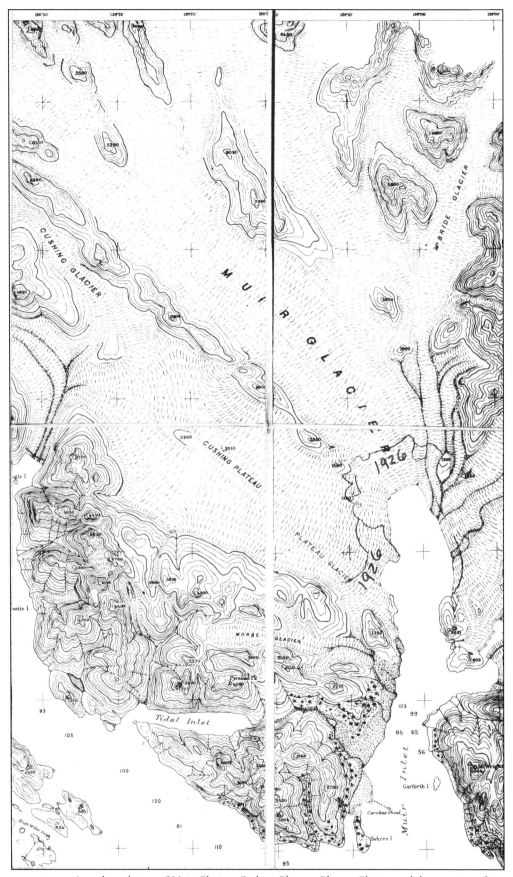

MAP 17. An enlarged view of Muir Glacier, Cushing Plateau, Plateau Glacier, and the terminus of Casement Glacier as shown on the map I took with me in 1926, the 1916 USC&GS map 8306 (see map 9 for the complete map). Camp Muir is identified but Goose Cove, where we stayed in 1926, was still under ice when the 1907 surveys upon which this map was based were made. I sketched in the terminus of Plateau, Muir, and Casement Glaciers while we were there.

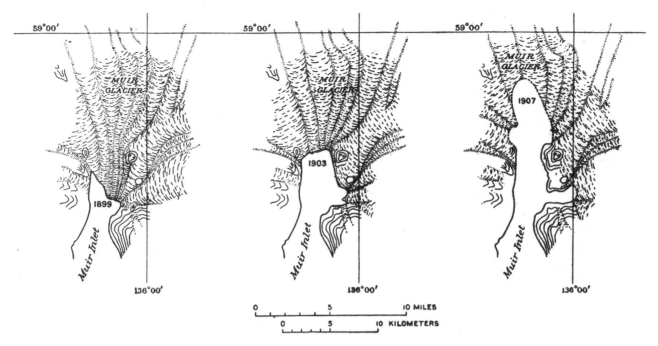

MAP 18. The retreat of Muir Glacier between 1899, as recorded by Henry Gannett of the Harriman Expedition, and 1907 when surveyed by Fremont Morse (1908:76–78) and Otto Klotz (1907: 419–421) of the International Boundary Commission. The 1903 position was mapped by C. L. Andrews (1903: 441–445). These maps were put together by Lawrence Martin for *Guide Book No. 10, Excursion C8* of the International Geological Congress in 1913.

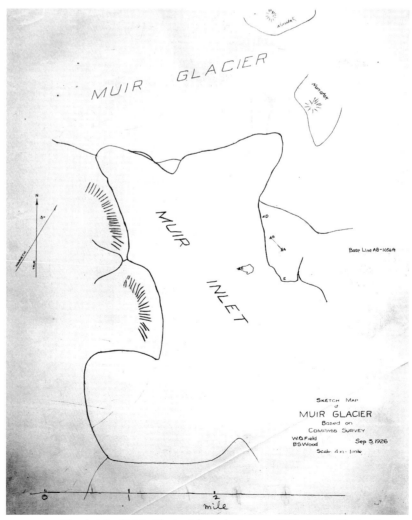

MAP 19. The sketch of Muir Glacier I drew during our 1926 visit to Glacier Bay.

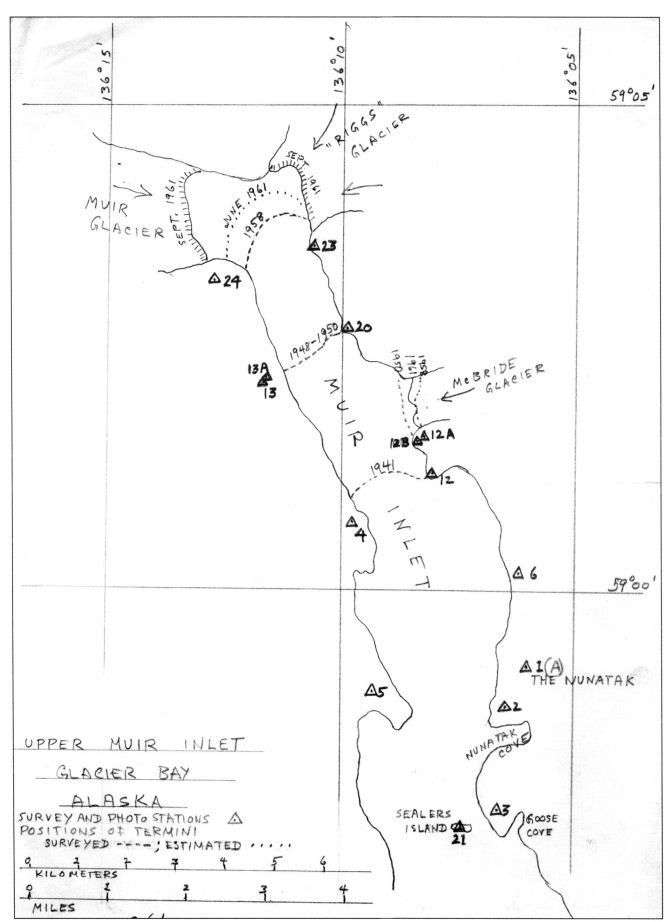

MAP 20. The map of Muir Inlet I made in 1961 showing the survey and photo stations and the recession of the terminus since 1941.

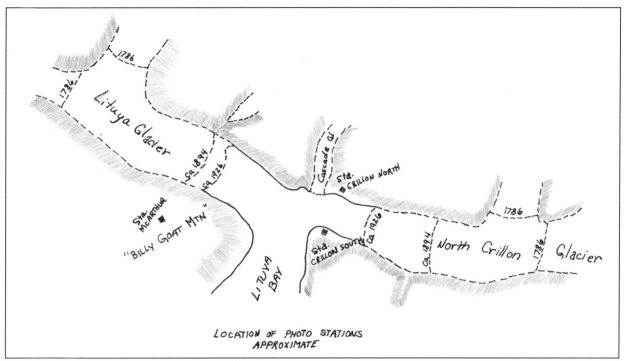

MAP 21. Lituya Bay and the approximate locations of the photo stations we occupied in 1926. Notice the advance of the glacier termini since 1786 when recorded by La Pérouse (see map 22).

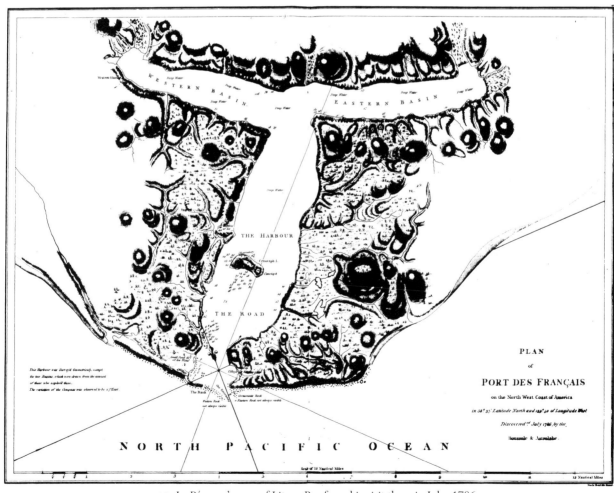

MAP 22. La Pérouse's map of Lituya Bay from his visit there in July, 1786.

SKETCH MAP
OF
GLACIER BAY
AND THE
FAIRWEATHER RANGE

SCALE IN MILES

▲ LOCATION OF PRINCIPAL PEAKS, (ELEVATIONS IN FEET).

⌐→ GLACIERS, (ARROWS INDICATE DIRECTION OF FLOW).

– – – ROUTE OF THE EXPEDITION, AUG. 13 – SEPT. 12, 1926.

······· FORMER POSITION OF ICE FRONTS (DATE INDICATED)

COMPILED FROM MAPS OF THE INTERNATIONAL BOUNDARY
COMMISSIONS, CHARTS OF THE U.S. COAST AND GEODETIC SURVEY,
AND PERSONAL OBSERVATIONS MADE DURING THE SUMMER OF 1926.

W. Osgood Field, 1926.

MAP 23. The map from my report on our 1926 trip to Glacier Bay. The route of our expedition is shown, as well as all the glacier terminus positions as we found them in 1926.

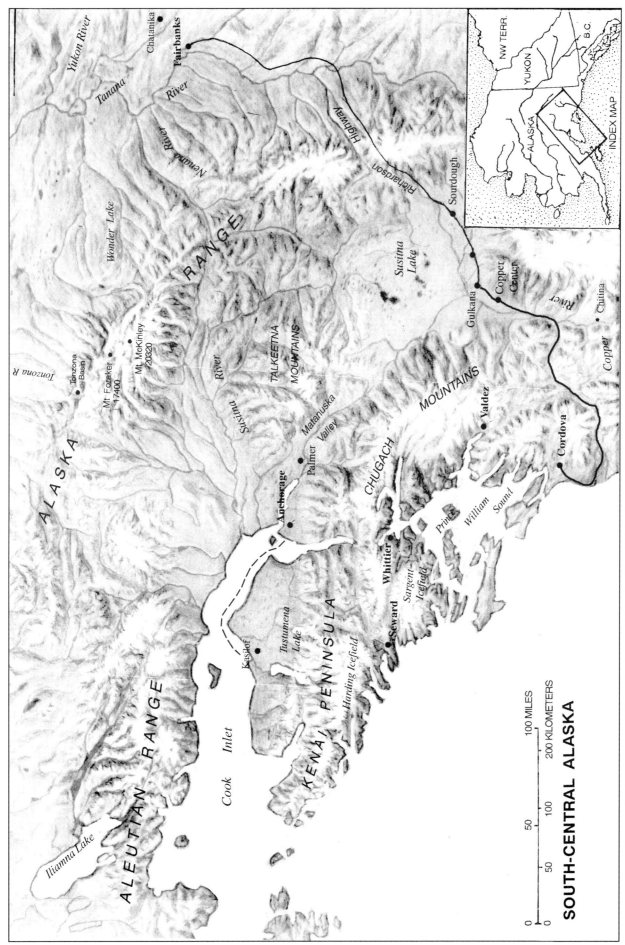

MAP 24. The overland route we took from Cordova to Fairbanks, from Fairbanks to the Tonzona Basin, and the boat route from Kasilof to Anchorage. © D. Molenaar 1990.

SOUTH-CENTRAL ALASKA

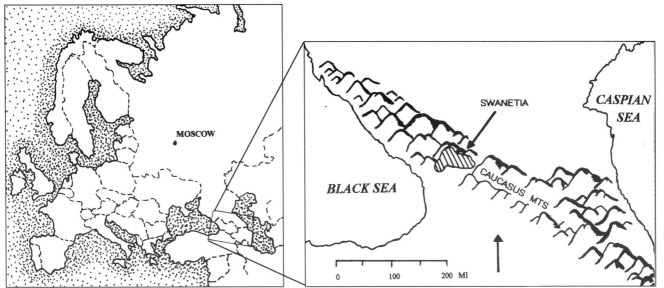

MAP 25. Swanetia is located in the Caucasus Mountains between the Black and Caspian Seas.

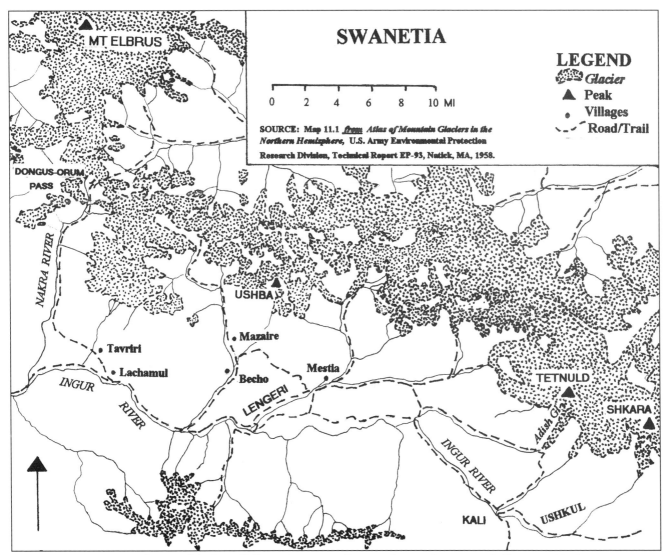

MAP 26. A close-up of the Ingur Valley in the western Caucasus showing the towns and villages we visited.

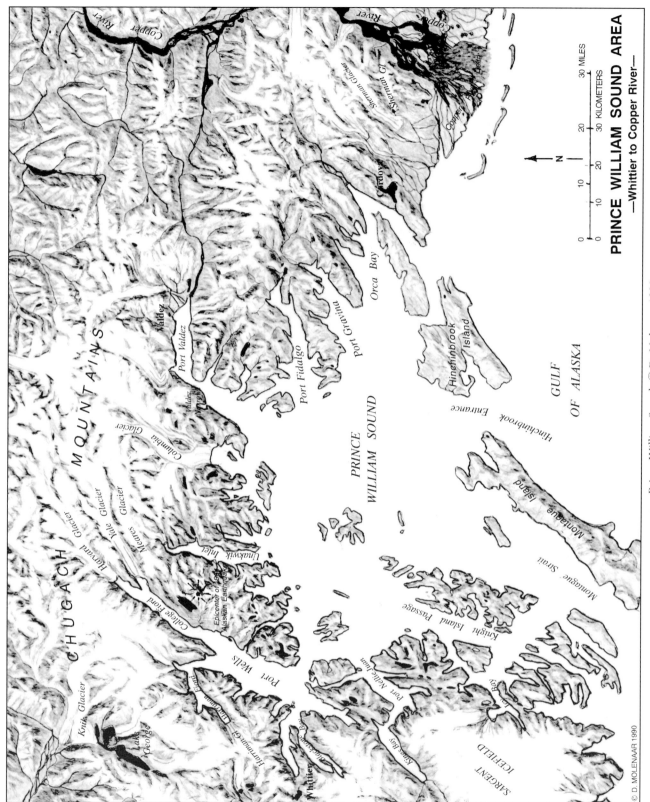

MAP 27. Prince William Sound. © D. Molenaar.

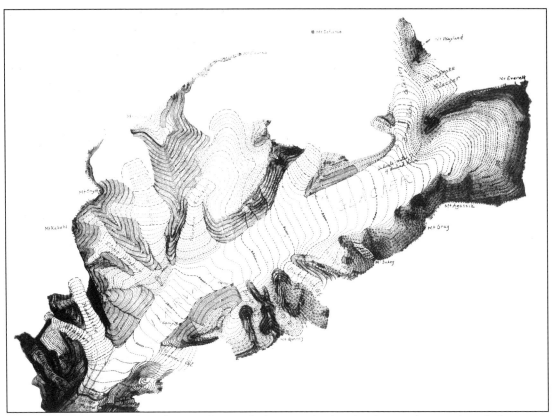

MAP 28. Dora Keen drew this map of Harvard Glacier in 1917 based on her trip there in 1915. [Scale is approximately 1 mile to 1 inch. Contour interval 100 feet.]

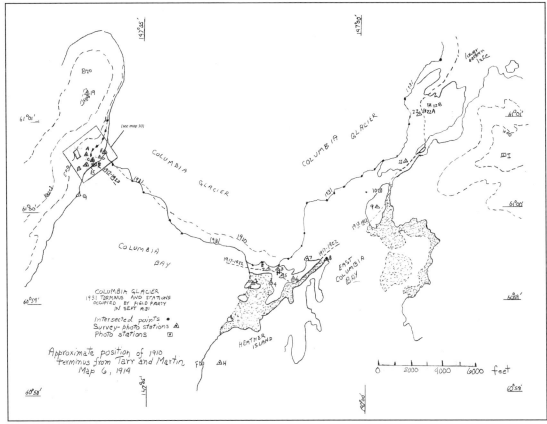

MAP 29. A map of Columbia Glacier I drew in 1986 based on my trip there in 1931. The lines labeled 1917–1922 mark the recent advance of Columbia Glacier that we established during that trip.

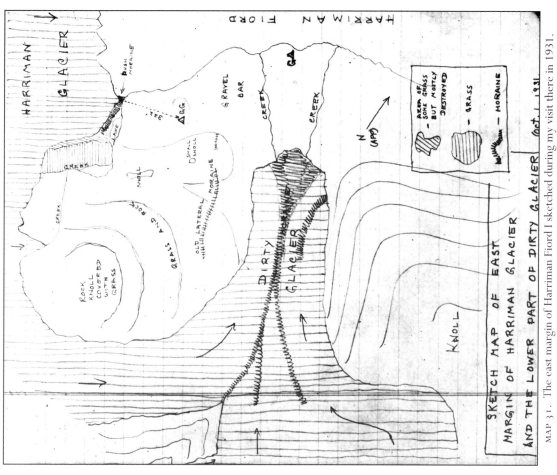

MAP 31. The east margin of Harriman Fiord I sketched during my visit there in 1931.

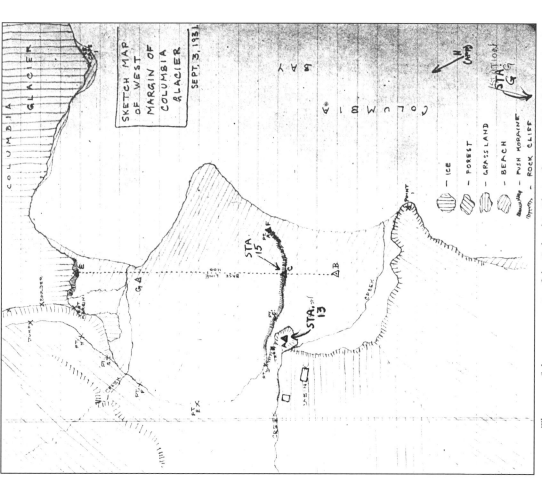

MAP 30. The map of the west margin of the Columbia Glacier from my 1931 field notebook. [Based on other maps in Bill's collection, the point below Sta. E labeled G should probably be labeled D. Sta. G is off the map, as the arrow in the lower right corner indicates. For the exact location of Sta. G, see Map 29. *Ed.*]

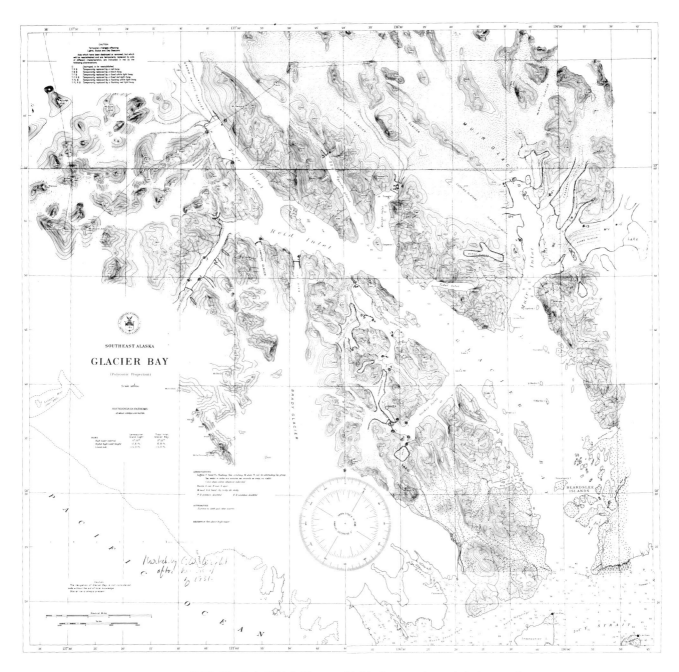

MAP 32. C.W. Wright's 1931 field map with his photo station and glacier notations.
(Original in W.O. Field Collection.)

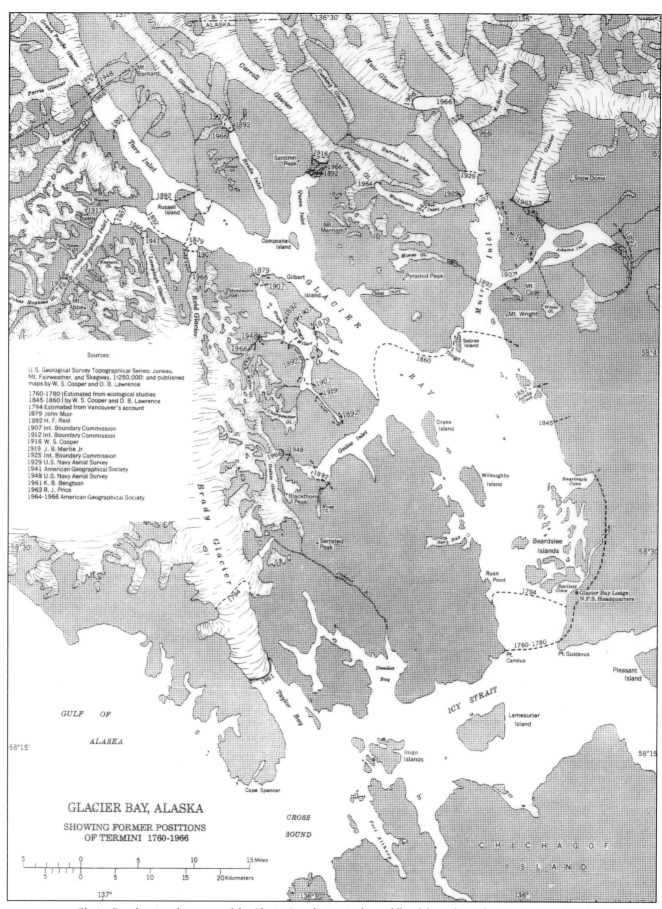

GLACIER BAY, ALASKA

SHOWING FORMER POSITIONS
OF TERMINI 1760-1966

Sources:

U.S. Geological Survey Topographical Series: Juneau,
Mt. Fairweather, and Skagway, 1:250,000; and published
maps by W. S. Cooper and D. B. Lawrence.

1760-1780 } Estimated from ecological studies
1845-1860 } by W. S. Cooper and D. B. Lawrence
1794 Estimated from Vancouver's account
1879 John Muir
1892 H. F. Reid
1907 Int. Boundary Commission
1912 Int. Boundary Commission
1916 W. S. Cooper
1919 J. B. Mertie Jr.
1925 Int. Boundary Commission
1929 U.S. Navy Aerial Survey
1941 American Geographical Society
1948 U.S. Navy Aerial Survey
1961 K. B. Bengtson
1963 R. J. Price
1964-1966 American Geographical Society

MAP 33. Glacier Bay showing the extent of the Glacier Bay glaciers in the middle of the eighteenth century. Note that the entire
bay was filled with ice. [This map was produced by the AGS under Bill's direction, probably for a slide presentation
or unpublished manuscript.]

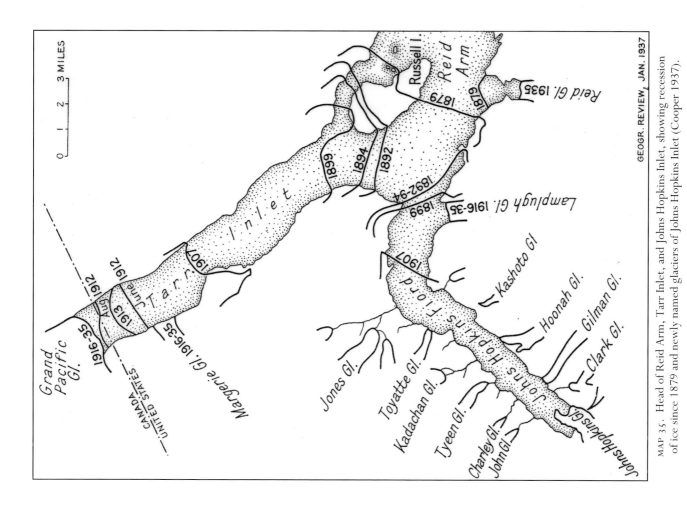

MAP 35. Head of Reid Arm, Tarr Inlet, and Johns Hopkins Inlet, showing recession of ice since 1879 and newly named glaciers of Johns Hopkins Inlet (Cooper 1937).

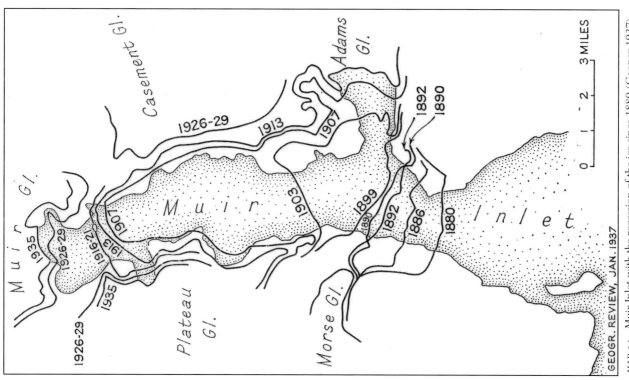

MAP 34. Muir Inlet with the recession of the ice since 1880 (Cooper 1937).

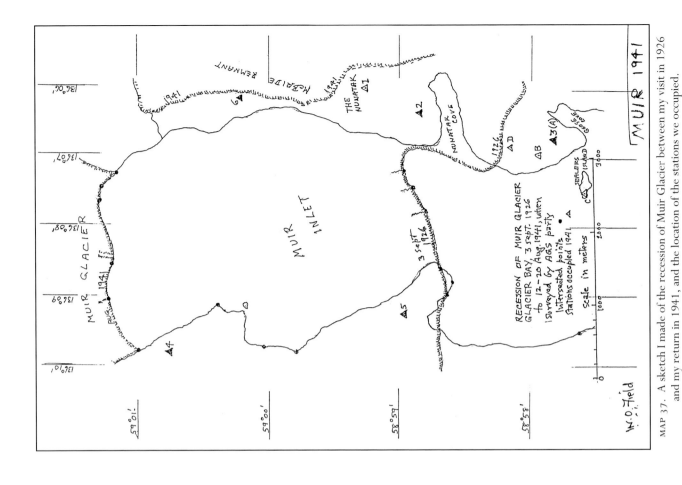

MAP 37. A sketch I made of the recession of Muir Glacier between my visit in 1926 and my return in 1941, and the location of the stations we occupied.

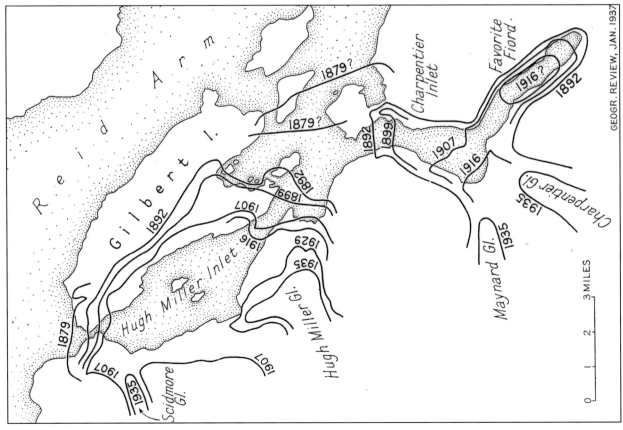

MAP 36. Hugh Miller Inlet and vicinity, illustrating recession of glaciers since 1879 (Cooper 1937).

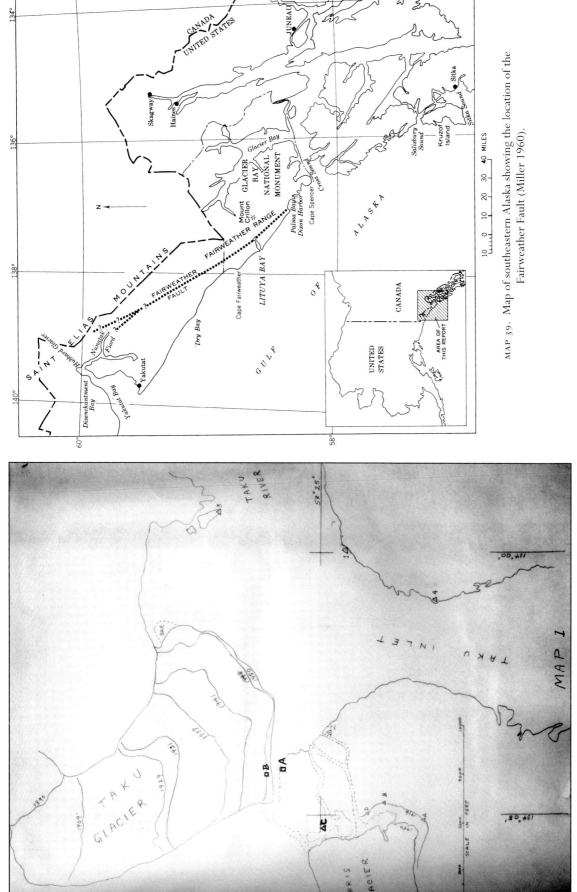

MAP 39. Map of southeastern Alaska showing the location of the Fairweather Fault (Miller 1960).

MAP 38. A sketch map prepared in 1955 by the Department of Exploration and Field Research for the Juneau Ice Field Research Project.

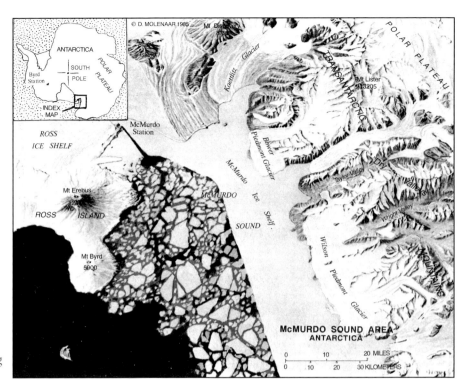

MAP 40. Antarctica showing its main
features and the places we visited during
our visit in 1957.

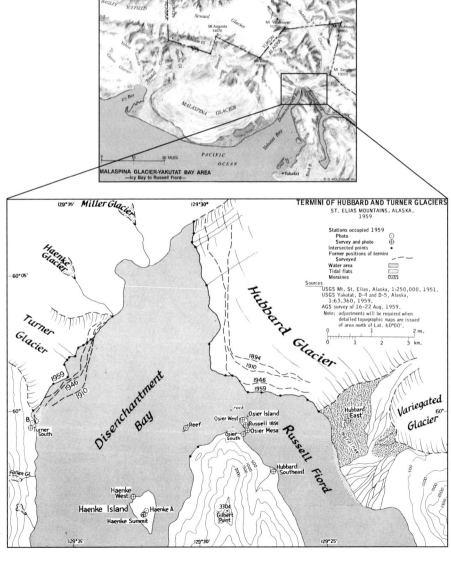

MAP 41. Disenchantment Bay and Hubbard
Glacier that we first visited in 1959.

APPENDIX A:
WILLIAM O. FIELD BIOGRAPHICAL OUTLINE

Jan 30, 1904	Born in New York City.
Jun 1922	Graduated Hotchkiss School.
Jul 1924	First Ascent, South Twin, Canadian Rockies.
Jun 1926	B.S. Geology, Harvard University.
1926	First scientific trip to Glacier Bay.
Apr 4, 1929	Married Alice Steelman Withrow.
Dec 30, 1930	Daughter Diana Sloane Field born.
Sep–Oct 1931	First scientific trip to Prince William Sound.
1931–57	Member, Committee on Glaciers, Section of Hydrology, American Geophysical Union (AGU).
Jun 13, 1940	Son John Osgood Field born.
Dec 2, 1940	Joined staff of the American Geographical Society (AGS) as a research associate.
Oct 1942– Jan 1946	World War II service as photographic officer in U.S. Army Signal Corps. Assignments included: Training Film Production Laboratory, Wright Field; Signal Corps Photographic Center, NY; Air Force Cold Weather Test Detachment in Alaska, winter 1943; Signal Photo Company in India and Burma, 1944–45. Attained rank of major at severance. Received Bronze Star Medal.
1947	Appointed head of the Department of Exploration and Field Research at the AGS.
1948–54	Chairman, Committee on Glaciers, Section of Hydrology, American Geophysical Union.
1954	Reporter on glaciology of the U.S. National Committee for the IGY (USNC/IGY).
1955–58	Chairman of the Technical Panel for Glaciology of the USNC/IGY. Technical Panel became the Glaciology Panel of the Committee on Polar Research (CPR)*, National Academy of Science/ National Research Council (NAS/ NRC) at the close of the IGY, 1958.
1954–57	Member, Technical Panel on Seismology and Gravity, USNC/ IGY.
1954–58	Consultant for the USNC/IGY Antarctic Committee.

1957	Director, World Data Center A: Glaciology.
1959–60	Member, Panel on Heat and Water, CPR/NAS.
1959	Member, Special Committee on IGY Geographic Names in the Antarctic, NAS.
1960	Member, Panel on Arctic Research, Third Arctic Planning Session, Geophysics Research Directorate, Air Force Cambridge Research Labs.
1960–63	Vice-President of Commission of Snow and Ice, International Association of Scientific Hydrology, International Union of Geodesy and Geophysics (IUGG).
1961–62	Committee of the British Glaciological Society.
1962–64	Vice-President of the International Glaciological Society (successor of the British Glaciological Society).
Aug 8, 1963	Married Mary Losey.
1964	Member of the Hydrology Panel of the NAS/NRC Committee of the Alaska Earthquake.
1969	Retired from the American Geographical Society.
1975	Publication of *Mountain Glaciers of the Northern Hemisphere*.
Jun 16, 1994	Died at his home in Great Barrington, MA.

Memberships

Fellow of the American Geographical Society

Fellow of the Geological Society of America

Fellow of the Arctic Institute of North America

International Glaciological Society (1970), honorary member

Alpine Club of Canada, member

American Geographical Society (1941), councilor and vice-president

American Alpine Club, councilor 1933–38, 1948– 49, eastern vice-president 1951–52

American Geophysical Union

Appalachian Mountain Club

Association of American Geographers

Camp Fire Club of America, secretary of conservation committee

Century Association

Council on Foreign Relations

Explorers Club

Harvard Club of New York

Harvard Mountaineering Club

Harvard Travelers Club

Himalayan Club

International Union of Geodesy and Geophysics

New York Academy of Sciences

Honors

1945: U.S. Army Bronze Star Medal

1961: Honorary D.Sc. degree, University of Alaska

1966: Antarctic Service Medal from NAS/NRC: "in recognition of your contributions to U.S. expeditions in Antarctica during the International Geophysical Year." Larry Gould, Director, U.S. IGY Antarctic Program

1969: Explorers Medal of the Explorers Club

1969: Charles P. Daly Medal of the AGS: "in recognition of your many contributions to the science of geography in the field of glaciology and for the conduct of your expeditions in Alaska and other areas." Serge A. Korff, President, AGS

1969: Glaciology Panel, CPR, NAS/NRC**

1978: Busk Medal of the Royal Geographical Society: "for your contributions to glaciological research and exploration in the northern hemisphere." John Hemming, Director and Secretary

1983: Harvard Travelers Club Medal

1983: Seligman Crystal of the International Glaciological Society: [The award is] "presented…to one who has contributed to glaciology in a unique way so that the subject is now significantly enriched as a result of that contribution. [Your nomination] was based…on your pioneering work on the variations of Alaska glaciers, your recognition and description of glacier surges, your cataloging of the glaciers of the world, your leadership in the development of U.S. glaciological programs in a worldwide context, and your influence in bringing others into glaciology in the days before U.S. glaciology gained its own momentum." Charles Swithinbank, President

Mountaineering and Climbing Record

1920: Canadian Rockies: Ptarmigan Peak

1923: Alps: Rimpfischhorn, Lyskamm, Matterhorn, Zinal Rothorn, Dent Blanche, Dufourspitze, Monte Rosa, Zumsteinspitz

1924: Canadian Rockies: 1st ascent unnamed peak on Columbia Ice Field (10,800 ft); 1st ascent South Twin Peak; 2nd ascent North Twin Peak; Mt. Castleguard; 3rd ascent Mt. Columbia (1st via southern ridge); 1st ascent unnamed peak (10,000 ft) on Mistya River; 1st ascent Mt. Patterson; 1st ascent Mt. Outram

1925: Canadian Rockies: 1st ascent unnamed peak (10,365 ft) near Pipestone Pass; 2nd ascent Mt. Clearwater; 1st ascent unnamed peak in Clearwater area

1925: Kenai Peninsula: ascent of small peak near Trail Glacier

1927: Canadian Rockies: Mt. Victoria

1928: Alps: peak near Innsbruck

1929: Alps: Wellenkuppe, Zermatt glaciers

1929: Caucasus: 2 trips to Swanetia in the Central Caucasus

* The Committee on Polar Research became the Polar Research Board and the Glaciology Panel is now PRB's Committee on Glaciology.

** Although not an award, Bill was commended for his work on the Glaciology Panel of the Committee on Polar Research, NAS/NRC. Following is an excerpt from the letter written on April 24, 1969, from A.L. Washburn, Chairman Glaciology Panel to Larry M. Gould, Chairman of the Committee on Polar Research expressing that commendation:

The work of the Academy's committees and panels is born willingly by scientists who are devoted to the best interests of their field and who accordingly endeavor to provide the necessary advice to the government, the scientific community, and the public at large. Often the work is extremely time consuming, with the lion's share of the task falling on the Chairman. This has long been an accepted fact of the Academy-National Research Council.

However, Bill Field's contributions while serving as Chairman of the Glaciology Panel have been so exceptional that the Panel unanimously and most enthusiastically wishes to commend Bill for his many services to glaciology in North America, to international cooperation in glaciology, and to the Glaciology Panel itself. With unswerving diligence and loyalty he has gently but firmly guided the Panel for these many years, starting with its inception as part of the IGY program, and continuing through post-IGY planning to the present day. Glaciology has now come of age as a widely recognized science, and we are proud to have had a chairman like Bill whose selfless leadership has contributed so significantly to this development. We honor him not only professionally but also as a person and a friend.

APPENDIX B: PUBLICATIONS OF WILLIAM O. FIELD

Mountaineering on the Columbia Icefield, 1924. 1925. *Bulletin of the Appalachia Mountain Club* 18, no. 12: 144–154 (June).

The Fairweather Range: Mountaineering and Glacier Studies. 1926. *Bulletin of the Appalachia Mountain Club* 20, no. 4: 460–472 (December).

In Search of Mount Clearwater. 1927. *Harvard Mountaineering* 1, no. 1: 5–11.

Unclimbed Peaks in the Alaska Range. 1928. *Harvard Mountaineering* 1, no. 2: 37–46.

Travel and Mountaineering in the Caucasus. 1930. *American Alpine Journal* 1, no. 2: 167–173.

Glaciers of the Northern Part of Prince William Sound, Alaska. 1932. *Geographical Review* 22, no. 3: 361–388.

The Mountains and Glaciers of Prince William Sound, Alaska. 1932. *American Alpine Journal* 1, no. 4: 445–458.

Glacier Studies. 1934. *American Alpine Journal* 2, no. 2: 279–282.

Alaskan Glacier Expedition, 1935. 1936. *American Alpine Journal* 2, no. 4: 548–551.

Observations on Alaskan Coastal Glaciers in 1935. 1937. *Geographical Review* 27, no. 1: 63–81.

(With W.S. Cooper.) Glacial Studies in Southern Alaska, 1935. *Annals of the Association of American Geographers* 26, no. 1: 44–45.

The Soviet Arctic. 1937. *Research Bulletin on the Soviet Union* 2, no. 2: 11–21.

The Kola Peninsula: Gibraltar of the Western Arctic. 1938. *American Quarterly on the Soviet Union* 1, no. 2: 3–21.

The North Pole Station Research Bulletin on the Soviet Union. 1938. Vol. 3, no. 4: 13, 16.

The International Struggle for Transcaucasia. 1939. *American Quarterly on the Soviet Union* 2, nos. 2–3: 21–44.

A Gazetteer of Alaskan Glaciers. 1941. *Transactions of the American Geophysical Union, Part III*: 796–799.

Glacier Studies in Alaska, 1941. 1942. *Geographical Review* 32, no. 1: 154–155.

Glacier Recession in Muir Inlet, Glacier Bay, Alaska. 1947. *Geographical Review* 37, no. 3: 369–399.

The Variation of Alaskan Glaciers, 1935–1947. 1948. *Procés-Verbaux*. Union Géodésique et Géophysique Internationale, Assemblée Générale d'Oslo: 227–282.

Glacier Observations in the Canadian Rockies, 1948. 1949. *Canadian Alpine Journal* 32: 99–114.

(With M.M. Miller.) Juneau Ice Field Research Project. 1950. *Geographical Review* 40, no. 2: 179–190.

Glaciological Research in Alaska. 1950. In *Selected Papers of the Alaskan Science Conference*. Washington, D.C.

Bibliography of Glaciological Research in Alaska. 1950. In *Alaskan Science Conference*, University of Alaska Fairbanks.

Problems of the Arctic: Glaciology, Glaciers, and the Arctic. 1950. Johns Hopkins Seminar (March).

Report on North American Glaciers, 1951. 1953. Union Géodésique et Géophysique Internationale, Assemblée Générale de Bruxelles 1: 120–128.

(With M.M. Miller.) Exploring the Juneau Ice Cap 1951. *Research Reviews* (April): 7–15.

Recent Glacier Variations in Glacier Bay, Alaska. 1951. *Transactions of the American Geophysical Union* 32: 331.

Studies of the Taku Glacier, Alaska. 1951. *Journal of Geology* 59: 622–623.

(With C.J. Heusser.) Glaciers: Historians of Climate. 1952. *Geographical Review* 42, no. 3: 337–345.

(With C.J. Heusser.) Glacier and Botanical Studies in the Canadian Rockies, 1953. 1954. *Canadian Alpine Journal* 37: 128–140.

Notes on the Advance of Taku Glacier. 1954. *Geographical Review* 44: 236–239.

Glaciers. 1955. *Scientific American* 193, no. 3: 84–92.

(With John K. Wright.) Lawrence Martin (obituary). 1955. *Geographical Review* 45: 587–588.

Some Aspects of Glaciers and Glaciology in the Dynamic North. 1956. *U.S. Dept. of Navy Technical Assistant for Polar Projects*, bk. 1, no. 8.

Flight to the South Pole. 1958. *Quad* 12, no. 2: 4–6, 8.

(Directed preparation.) *Geographic Study of Mountain Glaciation in the Northern Hemisphere*. 1958. American Geographical Society for the U.S. Army Quartermaster Research and Development Command, 10 parts in 3 vols.

Observations of Glacier Behavior in Southern Alaska, 1957. 1958. *IGY Glaciological Report Series*, no. 1, group 11.

Polar and Mountain Glaciers Under Study. 1959. *New York Herald Tribune*, January 11 (special supplement on IGY).

Antarctic Glaciology. 1961. *Science in Antarctica*. Part II, Publication 878, National Academy of Sciences, National Research Council.

(With A. Crary and M.F. Meier.) The United States Glaciological Researches During the

International Geophysical Year. 1962. *Journal of Glaciology* 4, no. 31.

Observations of Glacier Variations in Glacier Bay, Southeastern Alaska, 1958 and 1961. 1964. Preliminary Report (June).

(With R.H. Ragle and J.E. Sater.) Effects of the 1964 Alaskan Earthquake on Glaciers and Related Features. 1965. Arctic Institute of North America Research Paper no. 32 (April).

Mapping Glacier Termini in Southern Alaska, 1931–1964. 1966. *Canadian Journal of Earth Sciences* 3, no. 6, paper no. 10: 819–25.

Avalanches Caused by the Alaska Earthquake of March 1964. 1966. Publication no. 69 of the I.A.S.H., 326–331.

Profile: W.O. Field. 1966. *ICE*, no. 22.

Glaciology in the Arctic. 1967. *Transactions of the American Geophysical Union.* Committee on Polar Research, W.O. Field, Chairman.

Variations of Glaciers. 1968. *UBIQUE* 1, no. 1.

AGS Party in Alaska to Observe Sherman Landslide. 1968. *UBIQUE* 1, no. 3.

The Effect of Previous Earthquakes on Glaciers. 1968. In *The Great Alaska Earthquake of 1964*, vol. 3, *Hydrology*, pp. 252-265. National Academy of Sciences Publication 1603, Washington, D.C.

Effects on Glaciers, Introduction. 1968. In *The Great Alaska Earthquake of 1964*, vol. 3, *Hydrology*, pp. 249–251. National Academy of Sciences Publication 1603, Washington, D.C.

(With S. J. Tuthill and L. Clayton.) Postearthquake Studies at Sherman and Sheridan Glaciers. 1968. In *The Great Alaska Earthquake of 1964*, vol. 3,

Hydrology, pp. 318–328. National Academy of Sciences Publication 1603, Washington, D.C.

Current Observations on Three Surges in Glacier Bay, Alaska, 1965–1968. 1969. *Canadian Journal of Earth Sciences* 6, no. 4: 821–839.

(With E.V. Horvath.) References to Glacier Surges in North America. 1969. *Canadian Journal of Earth Sciences* 6, no. 4.

(Ed.) *Mountain Glaciers of the Northern Hemisphere.* 1975. 3 vols. Cold Regions Research and Engineering Laboratory, U.S. Army Corps of Engineers, Hanover, NH.

Observations of Glacier Variations in Glacier Bay National Monument. 1979. In *Proceedings of the First Conference on Scientific Research in the National Parks*, New Orleans, U.S. Dept of Interior (Nov. 9–12).

Developments in Wachusett Inlet and at Plateau Glacier. 1987. In J. Anderson et al. eds., *Observed Processes of Glacial Deposition in Glacier Bay, Alaska*, 41–46. Byrd Polar Research Center Miscellaneous Publication 236, Ohio State University, Columbus.

Glaciers of Alaska and Adjacent Yukon Territory and British Columbia. 1990. *American Alpine Journal* 32, no. 64: 79–149.

(With M. Sturm, D.K. Hall, and C.S. Benson.) Non-Climatic Control of Glacier-Terminus Fluctuations in the Wrangell and Chugach Mountains, Alaska, U.S.A. 1991. *Journal of Glaciology* 37, no. 127: 348–356.

(With D.K. Hall and C.S. Benson.) Changes of Glaciers in Glacier Bay, Alaska, Using Ground and Satellite Measurements. 1995. *Physical Geography* 16, no. 1: 27–41.

APPENDIX C: TRIP INFORMATION

Note: Information on members of the party and vessel(s) used for each trip was obtained from captions in the WOF trip photo albums or from Field's diaries. Occasionally neither source contained the information.

1926 Glacier Bay

DATES: 6 August–3 September

PARTY: William O. Field, Andy Taylor, Roscoe ('Rocky') Bonsal, Percy Pond, Ben Wood

VESSEL: M/V *Eurus*; Paul Kegal, skipper

GLACIERS/INLETS VISITED: East & West Twin Gls, Hole-in-the-Wall Gl, Taku Gl, Norris Gl, Muir Inlet, Carroll Gl, Rendu Inlet, Tarr Inlet, Johns Hopkins Inlet, Lamplugh Gl, Reid Gl, Scidmore Gl, Hugh Miller Gl, Charpentier Inlet, Geikie Inlet, Wood Gl, Finger Gl, LaPerouse Gl, Lituya Bay

1931 Prince William Sound

DATES: 1 September–14 October

Party: William O. Field, Andy Taylor, Sherman Pratt, Jack Wood

Vessel: M/V *Virginia*; Rex and Ronney Hancock, owners/operators; Charlie Lilygren, asst pilot

GLACIERS/INLETS VISITED: Childs Gl, Miles Gl, Valdez Gl, Anderson Gl, Shoup Gl, Columbia Gl, Unakwik Inlet, College Fiord, Barry Arm, Harriman Fiord

1935 Glacier Bay and Prince William Sound

DATES: 1 August–9 October

PARTY: William O. Field, Bob Stix, William S. Cooper, Russ Dow

VESSEL: Glacier Bay: M/V *Yakobi*; Tom Smith, skipper; Ed, cook. Prince William Sound: M/V *Fidelity*; Les Newton, cook; Frank Granite, engineer.

GLACIERS/INLETS VISITED: Sumdum Gl, Dawes Gl, N. Dawes Gl, Sawyer Gl, S. Sawyer Gl, Davidson Gl, Rendu Inlet, Tarr Inlet, Johns Hopkins Inlet, Lamplugh Gl, Reid Gl, Hugh Miller Gl, Charpentier Inlet, Geikie Inlet, Brady Gl, LaPerouse Gl, Childs Gl, Miles Gl, Allen Gl, Valdez Gl, Shoup Gl, Columbia Gl, Meares Gl, College Fiord, Barry Arm, Harriman Fiord, Blackstone Bay, Tebenkof Gl, Cochrane Bay, Kings Bay, Nellie Juan Gl, Ultramarine Gl, Icy Bay (in PWS)

1941 Coast Mountains and Glacier Bay

DATES: 8 August–20 September

PARTY: William O. Field, A. Tony Ladd, Don Lawrence, Maynard M. Miller

VESSEL: Coast Mountains: U.S.F.S. *Forester*. Glacier Bay: M/V *Treva C.*

GLACIERS/INLETS VISITED: Sumdum Gl, Patterson Gl, Baird Gl, Dawes Gl, N. Dawes Gl, Brown Gl, Sawyer Gl, S. Sawyer Gl, East & West Twin Gls, Taku Inlet, Muir Inlet, Wachusett Inlet, Adams Inlet, Queen Inlet, Rendu Inlet, Tarr Inlet, Johns Hopkins Inlet, Lamplugh Gl, Reid Gl, Hugh Miller Gl, Charpentier Inlet, Geikie Inlet

1948 Canadian Rockies

DATES: 22 July–8 August

Party: William O. Field, Diana Field, Frank Wells (outfitter Jasper), George Wells (assistant), Charley Pearson (assistant)

GLACIERS/STATIONS VISITED: Columbia Gl (Sta. 1, Sta. 2, Sta. 3, Sta. A, Palmer 1924 site, icefall, terminus), Athabaska Gl (Sta. 1, Sta. 2, Wilcox Ridge [1922 site]), Saskatchewan Gl (Sta. 21, Sta. 22, Thorington 1923 site, Sta. C [became Sta. 23], Sta. A [became Sta. 26], below Castleguard Sta. 1 [Castleguard Pass]), Habel Creek and Horseshoe Gl (Sta. A), Stutfield Gl (from hwy), Dome Gl, Victoria Gl

1949 Canadian Rockies

DATES: 22 July–26 August

PARTY: William O. Field, Diana Field, Paul Livingston, Len Jack, Frank Wells

GLACIERS/STATIONS VISITED: Toronto Gl, Wales Gl, Columbia Gl (Sta. South Bar, Sta. South Slide, Sta. Trees, Sta. Boulder, Sta. 1, Sta. 2, Sta. 3, Sta. 4, Sta. 5, Sta. Palmer SW, Sta. 6, Sta. 11, Sta. 12, Sta. 13, terminal lake), Athabaska Gl (Sta. 2, Sta. 5, Sta. 6, Sta. Schaeffer 1907, terminus, Nigel Peak), Saskatchewan Gl (Sta. Mitchell 1912, Sta. 20 [Parker Ridge], Sta. 21 [Parker Ridge], Sta. 22 [Parker Ridge], Sta. 23, Sta. 25, Sta. 26), Castleguard Gl (Sta. 1 on Castleguard Pass, Sta. 2 [old U.S. Army camp on Sas Gl], Sta. 4, Sta. 5, Sta. 6, IBS Sta. 86 [Interprovincial BS], Sta. 7, Sta. 8, Sta. 9, Sta. 10, Sta. 11, Sta. 12, Sta. 13, Sta. 14, Sta. Bryce, Sta. 15, Sta. A of Terrace Gl No. 4), N. Rice Gl (Sta. A, base of Mt. Terrace, outer moraine of Castleguard Gl No. 1, Castleguard Gl No. 3), Alexandra Gl (Sta. A, Sta. B), Bow Gl (from hwy), Crowfoot Gl (from hwy)

1950 Coast Mountains and Glacier Bay

DATES: 27 July–7 September

Party: William O. Field, Don B. Lawrence, Diana Field, E.G. Lawrence, Paul Livingston

VESSEL: Coast Mountains: U.S.F.S. *Ranger 10*; Glacier Bay: M/V *Galavanter*

GLACIERS/INLETS VISITED: Dawes Gl, N. Dawes Gl, Sawyer Gl, S. Sawyer Gl, Hole-in-the-Wall Gl, Taku Gl, Norris Gl, Adams Inlet, Muir Inlet, Wachusett Inlet, Queen Inlet, Rendu Inlet, Tarr Inlet, Johns Hopkins Inlet, Lamplugh Gl, Reid Gl, Hugh Miller Gl, Geikie Inlet

1953 Canadian Rockies

DATES: 13 July–12 September

PARTY: William O. Field, Cal Heusser, Howell Archard, Steve den Hertog, Sandy, Jim Simpson, Sr., Jim Simpson, Jr.

GLACIERS/STATIONS VISITED: Columbia Gl (Sta. Palmer Trees, Sta. Palmer Bar, Sta. Palmer Boulder, Sta. Slide, Sta. 1, Sta. 2, Sta. 3, Sta. 3A, Sta. 5, along S lateral moraine, Sta. 5A, Sta. 6, Sta. 13, Sta. West Knoll, ice fall), Dais Gl (Pt A, Pt B, Pt C), Berg Gl (slopes of Mumm Ridge, east margin, terminus), SE Lyell Gl (Walcott's 1919 site, above Glacier Lake, Sta. A, Sta. 1 [BS 1918 Sta. Lake Moraine], Sta. 2, base of Butte, Kothe's 1940 site, Thorington's 1930 site N side Gl River, Sta. 5), Freshfield Gl (Sta. 1, Sta. C [Palmer's 1922 site], Sta. 2, Thorington's 1922 erratic, Sta. 3, Sta. 4, Sta. 5, Sta. 6, Collie's 1897 erratic, Niverville Meadow [sites 1, 2, 4, 5, 6, 7]), Robson Gl (1922 terminus, Boundary Monument Robson Pass, Sampson 1931 site, Sta. 1, Sta. 2, Sta. 3, Sta. 5, Sta. 6, Sta. 7, Sta. 8, Sta. 9, Sta. 10, outer moraine, lower slopes Mumm Ridge, Pt Wheeler 1911 site, Coleman 1908 site), Saskatchewan Gl (Sta. 20, Sta. 21, Sta. 22, Sta. 25, Sta. 26, Sta. 27, Sta. 28), Athabaska Gl & Dome Gl (Sta. 1, Sta. 2, Sta. 3, Sta. 5, Sta. 6, Sta. 7, Sta. 8, Sta. 9, Dome Sta. A, W & SW lateral moraines, Sta. 10, Dome Gl lateral moraines, Thorington 1923 site, near Columbia Icefield Chalet, Wilcox Ridge), Hilda Gl (from hwy, outer moraines), Angel Gl (moraines, Sta. North), Peyto Gl (lookout above Peyto Lake, Wilcox site of 1896, Sta. 2, Sta. 4, Sta. 5, Sta. 5A, Sta. 7, Water Bureau [Sta. TP 47, Sta. 1, Sta. 6, Sta. 8]), Bow Gl (Sta. A, Sta. B, end moraines and trim lines), Yoho Gl (Habel 1897 site, Sherzer 1907 site 'End of the Trail,' Sta. 2, Vaux 1902 site, Sta. Rock 1 [Wheeler, 1906], Sta. Rock 1A, Sta. Rock 2 [Wheeler, 1908], near Vaux 1901 site, Sta. 1, Sta. 4 [Wheeler Sta. C]), Wenkchemna Gl

1957 Prince William Sound, Kenai, and Alaska Range

DATES: 4 July–8 September

PARTY: William O. Field, Leslie A. Viereck, Eleanor Viereck, Marion Millett, Robert J. Goodwin, Charles C. Morrison, Jack Major

VESSEL: M/V *Valiant Maid*; Ed Bilderback, skipper; Willie, deck hand

GLACIERS/INLETS VISITED: Childs Gl, Worthington Gl, Tazlina Gl, Nelchina Gl, Valdez Gl, Shoup Gl, Columbia Gl, Unakwik Inlet, College Fiord, Barry Arm, Harriman Fiord, Blackstone Bay, Tebenkof Gl, Kings Bay, Nellie Juan Gl, Utramarine Gl, Icy Bay (in PWS), Bainbridge Gl, Portage Gl, Trail Gl, Bartlett Gl, Spencer Gl, Deadman Gl, Bear Gl, Matanuska Gl, Kennecott Gl, Gulkana Gl, "West Gulkana," Muldrow Gl

1958 Coast Mountains and Glacier Bay

DATES: 26 July–12 September

PARTY: William O. Field, Marie T. Hatcher, Marion T. Millett, John Cornejo, Charles C. Morrison

VESSEL: Glacier Bay: NPS M/V *Nunatak II*, Ken Youmans, NPS skipper

GLACIERS/INLETS VISITED: Sumdum Gl, Dawes Gl, N. Dawes Gl, Sawyer Gl, S. Sawyer Gl, Hole-in-the-Wall Gl, Taku Gl, Norris Gl, Mendenhall Gl, Herbert Gl, Eagle Gl, Denver Gl, Laughton Gl, Ferebee Gl, Rainy Hollow Gl, Jarvis Gl, Tsirku River, Takhin River, Garrison Gl, Rainbow Gl, Davidson Gl, Muir Inlet, Wachusett Inlet, Carroll Gl, Rendu Inlet, Tarr Inlet, Johns Hopkins Inlet, Lamplugh Gl, Reid Gl, Scidmore Gl, Hugh Miller Gl, Charpentier Inlet, Geikie Inlet

1959

Dates: 31 July–3 September

Party: William O. Field, Richard Goldthwait, Marion T. Millett, George Burns, John O. Field, Barney (skipper), George Adams (first mate)

GLACIERS/INLETS VISITED: Hole-in-the-Wall Gl, Taku Gl, Norris Gl, Mendenhall Gl, Disenchantment Bay, Russell Fiord, Nunatak Fiord, Childs Gl, Miles Gl, Sheridan Gl, Portage Gl, Gulkana Gl, Muldrow Gl

1961 Glacier Bay and Prince William Sound

DATES: 9 August–18 September

PARTY: William O. Field, John O. Field, Marion T. Millett, D.B. Butts (GB), Lawrence Sardoni (PWS), Charles Peterson (PWS)

VESSEL: Glacier Bay: NPS M/V *Nunatak II*; Wendell Schneider, NPS skipper; Ken Youmans, deck hand. Prince William Sound: M/V *Valiant Maid;* Ed and Danny Bilderback, owners/ skipper

GLACIERS/INLETS VISITED: Muir Inlet, Wachusett Inlet, Carroll Gl, Rendu Inlet, Tarr Inlet, Johns Hopkins Inlet, Lamplugh Gl, Reid Gl, Hugh Miller Gl, Geikie Inlet, Worthington Gl, Childs Gl, Sheridan Gl, Valdez Gl, Shoup Gl, Columbia Gl, Meares Gl, College Fiord,

Harriman Fiord, Tebenkof Gl, Taylor Gl, Falling Gl, Nellie Juan Gl, Trail Gl, Bartlett Gl, Spencer Gl

1963 Canadian Rockies

DATES: 9–12 September

PARTY: William O. Field, Mary L. Field

GLACIERS/STATIONS VISITED: Athabaska and Dome Gls (Sta. 1, Sta. 2, Sta. 5, Sta. 6, hill behind Columbia Icefield Chalet, Sta. Wilcox South, Sta. Wilcox North, Sta. Wilcox North A, Sta. Wilcox North B) Saskatchewan Gl (Sta. 20, Sta. 22), Bow & Crowfoot Gls (from highway), Peyto Gl (from lookout above Peyto Lake)

1964 Glacier Bay and Prince William Sound

DATES: 13 August–1 October

PARTY: William O. Field, Richard P. Goldthwait, Kay Goldthwait, Peter Mapes, Marion T. Millett, Charles Janda, (GB)

VESSEL: Glacier Bay: NPS M/V *Nunatak II*; Jim Sanders, NPS skipper. Prince William Sound: M.S. *Valiant Maid*; Ed Bilderback, skipper; Franco, crew

GLACIERS/INLETS VISITED: Mendenhall Gl, Herbert Gl, Eagle Gl, Muir Inlet, Casement Gl, Wachusett Inlet, Carroll Gl, Rendu Inlet, Tarr Inlet, Johns Hopkins Inlet, Lamplugh Gl, Reid Gl, Hugh Miller Gl, Charpentier Inlet, Geikie Inlet, Sherman Gl, Sheridan Gl, Worthington Gl, Valdez Gl, Shoup Gl, Columbia Gl, Unakwik Inlet, College Fiord, Barry Arm, Harriman Fiord, Blackstone Bay, Tebenkof Gl, Kings Bay, Nellie Juan Gl, Ultramarine Gl, Bartlett Gl, Spencer Gl, Portage Gl

1965

DATES: 29 June–21 July

PARTY: William O. Field, Marion T. Millett, Lee Clayton

GLACIERS/INLETS VISITED: East and West Twin Gls, Hole-in-the-Wall Gl, Taku Gl, Norris Gl, Mendenhall Gl, Sherman Gl, Sheridan Gl

1966 Glacier Bay, Prince William Sound, Kenai

DATES: 11 August–21 September

PARTY: William O. Field, Marion T. Millett, Dave Bohn, Lynn Kinsman, Bob Howe (Supt. GB), Bob J. Goodwin (PWS)

VESSEL: Glacier Bay: NPS M/V *Nunatak II;* Jim Sanders, NPS skipper. Prince William Sound: M.S. *Citation*; Harry Richards

GLACIERS/INLETS VISITED: Muir Inlet, Wachusett Inlet, Carroll Gl, Rendu Inlet, Tarr Inlet, Johns Hopkins Inlet, Lamplugh Gl, Reid Gl, Hugh Miller Gl, Geikie Inlet, Sherman Gl, Sheridan Gl, Worthington Gl, Valdez Gl, Columbia Gl, College Fiord, Barry Arm,

Harriman Fiord, Blackstone Bay, Tebenkof Gl, Kings Bay, Nellie Juan Gl, Ultramarine Gl, Icy Bay (in PWS), Trail Gl, Bartlett Gl, Spencer Gl, Portage Gl

1967

DATES: 10–29 July

PARTY: William O. Field, Mary L. Field, George Haselton, Bob Howe (Supt. GB), Greg Streveler (GB)

VESSEL: Coast Mtns: M/V *Capella*; Bill and Helen Hixson, owners/skipper. Taku Inlet: Taku Lodge skiff; Dick Gregg, skipper. Glacier Bay: NPS M/V *Nunatak II*; Jim Sanders, NPS skipper

GLACIERS/INLETS VISITED: Baird Gl, Dawes Gl, Sawyer Gl, S. Sawyer Gl, Wright Gl, East & West Twin Gls, Hole-in-the-Wall Gl, Taku Gl, Norris Gl, Lemon Creek Gl, Mendenhall Gl, Muir Inlet, Wachusett Inlet, Carroll Gl, Rendu Inlet, Tarr Inlet, Johns Hopkins Inlet, Lamplugh Gl, Reid Gl, Steele Gl, Rusty (Fox) Gl, Jackal Gl, Donjek Gl

1968

DATES: 7–31 July

PARTY: William O. Field, Marion T. Millett, Joe Lakovitch, Bob Howe (Supt. GB), Russ Cahill (GB)

VESSEL: Taku Inlet: Taku Lodge skiff; Bill Bixby, skipper. Glacier Bay: NPS M/V *Nunatak II*; Jim Sanders, NPS skipper; Bill Meyers, deck hand

GLACIERS/INLETS VISITED: East & West Twin Gls, Hole-in-the-Wall Gl, Taku Gl, Norris Gl, Mendenhall Gl, Muir Inlet, Wachusett Inlet, Carroll Gl, Rendu Inlet, Tarr Inlet, Johns Hopkins Inlet, Lamplugh Gl, Reid Gl, Scidmore Gl, Hugh Miller Gl, Geikie Inlet, Childs Gl, Sherman Gl, Sheridan Gl

1971

DATES: 21 July–3 August

PARTY: William O. Field, Austin Post, Marion T. Millett, Bob Howe (Supt. GB)

VESSEL: Glacier Bay: NPS M/V *Nunatak III*; Jim Sanders, NPS skipper

GLACIERS/INLETS VISITED: Mendenhall Gl, Muir Inlet, Wachusett Inlet, Carroll Gl, Rendu Inlet, Tarr Inlet, Johns Hopkins Inlet, Lamplugh Gl, Reid Gl, Hugh Miller Gl, Geikie Inlet, Sherman Gl, Sheridan Gl, College Fiord, Barry Arm, Harriman Fiord, Portage Gl

1974

DATES: 30 June–12 August

PARTY: William O. Field, Richard H. Ragle, Dave Bohn, George Haselton, Bob Howe (Supt. GB), Marion T. Millett (PWS), Fred Joiner (PWS)

VESSEL: Glacier Bay: NPS M/V *Nunatak III*; Jim Sanders, NPS skipper; Gene Chaffin, deck hand. Kenai Mtns: M/V *Sea Hawk II*; Ed Shepard, owner(?)/skipper

GLACIERS/INLETS VISITED: Wright Gl, East and West Twin Gls, Hole-in-the-Wall Gl, Taku Gl, Norris Gl, Mendenhall Gl, Muir Inlet, Wachusett Inlet, Carroll Gl, Rendu Inlet, Tarr Inlet, Johns Hopkins Inlet, Lamplugh Gl, Reid Gl, Hugh Miller Gl, Charpentier Inlet, Geikie Inlet, Childs Gl, Sherman Gl, Sheridan Gl, Unakwik Inlet, College Fiord, Barry Arm, Harriman Fiord, Blackstone Bay, Tebenkof Gl, Kings Bay, Nellie Juan Gl, Trail Gl, Bartlett Gl, Spencer Gl

1976

DATES: 2 July–11 August

PARTY: William O. Field, John O. Field (GB), Bob Howe (Supt. GB, Ret.), Tom Ritter (Supt. GB), Rick Caulfield (GB), Mary L. Field (PWS), Richard Ragle (PWS)

VESSEL: Glacier Bay: NPS M/V *Nunatak III*; Jim Sanders, NPS skipper; Gene Chaffin, deck hand. Prince William Sound: M/V *Shamin*; Don and Pam Oldow

GLACIERS/INLETS VISITED: Mendenhall Gl, Muir Inlet, Wachusett Inlet, Rendu Inlet, Tarr Inlet, Johns Hopkins Inlet, Lamplugh Gl, Reid Gl, Scidmore Gl, Hugh Miller Gl, Charpentier Inlet, Geikie Inlet, Worthington Gl, Childs Gl, Sherman Gl, Sheridan Gl, Valdez Gl, Shoup Gl, Columbia Gl, Unakwik Inlet, College Fiord, Barry Arm, Harriman Fiord, Blackstone Bay, Kings Bay, Nellie Juan Gl, Ultramarine Gl

1982

All pictures of glaciers in Glacier Bay were taken from the boat; no photo/survey stations were occupied.

DATES: 17–23 June

PARTY: William O. Field, John O. Field, Bob Howe (Supt. GB, Ret.)

VESSEL: Glacier Bay: NPS M/V *Nunatak III*; Gene Chaffin, NPS skipper; Jim Luthy, deck hand

GLACIERS/INLETS VISITED: Wright, East & West Twin Gls, Hole-in-the-Wall Gl, Taku Gl, Norris Gl, Muir Inlet, Wachusett Inlet, Tarr Inlet, Johns Hopkins Inlet, Lamplugh Gl, Reid Gl, Rendu Gl

1983

While attending the First Glacier Bay Science Symposium, Bill took a one-day excursion on the NPS M/V *Nunatak III* to Muir Inlet on September 22.

1987

Bill took some aerial photographs during a brief stay in June.

APPENDIX D: ALASKA GLACIER PHOTOGRAPHERS
A Partial Listing of Terrestrial Photographers to 1935, Taken from Material by W.O. Field

ABC - Alaska Boundary Commission
DLS - Dominion Land Surveyor
Gl - glacier
IBC - International Boundary Commission
PWS - Prince William Sound
USC&GS - U.S. Coast and Geodetic Survey
USGS - U.S. Geological Survey

Note: If a photographer was known to be in only one specific area of Alaska, that area is noted in parentheses.

Photographer	Dates	Affiliation/Notes
Alexander, B.	1911	IBC (Disenchantment Bay)
Andrews, C.L.	1899, 1903, 09, 12, 13, 15, 16	U.S. Customs, Skagway
Andrews, E.	ca. 1910, 15	
Atwood, W.W.	pre-1930	USGS
Baldwin, Asa C.	1909, 10, 11, 12, 13	IBC
Baldwin, G. Clyde	1906, 07, 08, 09	IBC
Baldwin, Prentiss	1886, 90, 92	
Bigger, C.A.	1904, 05	IBC
Black, R. Clifford	1891	Possibly with Veazie Wilson
Blanchard	1923	(Muldrow Gl)
Bliss	1897	ABC
Brabazon, A.J.	1893, 94, 95, 1905, 06	DLS, ABC, IBC
Brabazon, Claude H.	1905, 06, 08	IBC
Brock, J.W.	1913	Int'l Geological Congress (Dept. of Mines)
Brooks, A.H.	1899, 1902, 12, 24	USGS
Brown, Webster	1910	USGS
Bryant, H.G.	1897	Philadelphia Geographical Society (Disenchantment Bay)
Buddington, A.F.	1922, 23, 24, 25	USGS
Capps, S.R.	1910, 17, 27, 28	USGS
Cantwell, G.G.	pre-1913	Valdez (PWS)
Carlin, W.E.	1909	Photos taken for the George W. Perkins party (PWS)
Case, W.M.	1903, 07, 10	Possibly with Draper
Case and Draper	1890s, 1903	
Chapin, Theo.	1911	USGS
Clunn, T.H.G.	1909	IBC
Cochrane	1909	
Cooper, W.S.	1916, 21, 29, 35	University of Minnesota
Craig, J.D.	1905, 06, 07, 08, 09, 10, 11, 12, 13, 20	IBC
Curtis, E.S.	1899	Harriman Expedition, official photographer
Davidson, George	1883	USC&GS
Davis, Trevor	1912, 14, 15, 16, 19, ca. 1922	Commercial photographer, Juneau
Dennis, T.C.	1911, 12, 13, 20	IBC
Dennis, W.M.	1910, 11, 12, 13, 14	IBC
Decoeli, E.T.	1908, 13	IBC
Dunann, C.D.	1911	On S.S. *Spokane*, possibly with Tarr & Martin
Eaton, D.W.	1904, 05, 06, 07, 08, 09 10, 11, 12, 13	IBC
Ellsworth	probably 1890s–early 1900s	USGS (PWS)
Field, William O.	1925, 26, 27, 31, 35	Personal

Photographer	Dates	Affiliation/Notes
Flemer, J.A.	1893, 94, 98, 1900, 04, 05	ABC, IBC
Gibbons, James L.	1893, 94	DLS, ABC
Gibbons, Jas G.	1893, 94	ABC
Gilbert, G.K.	1899	USGS (Harriman Expedition)
Gilpatrick, J.H.	1928	Photo shop, Sitka
Grant, U.S.	1908, 09	USGS (PWS)
Handy, Dora Keene	1914, 25	Personal (PWS)
Hatch, Laura	1913	Int'l Geological Congress
Higgins, D.F.	1908, 09	USGS (PWS)
Hill, Jesse	1908, 09, 10, 20	IBC
Hills, T.M.	1910, 13	Int'l Geological Congress
Hordern, Radcliffe	1905, 06, 07	IBC
Hunt, P.S.	pre-1913	Valdez (PWS)
Ives Heath A.	1920s	
Johnson, B.L.	1910, 11, 13, 14, 15, 16	USGS (PWS 1914–16)
Klotz, Otto J.	1889, 93, 94, 1906	DLS, DTS, ABC
Knopf, Adolph	1909, 10	USGS
LaRoche	1893, 94, 97	Commercial photographer, Seattle, WA
Lambart, H. Fred	1906, 07, 08, 09, 10, 11, 12, 13	IBC
Landes, K.K.		USGS (Knik Gl)
Latham, E.B.	1926	USC&GS (Steamer *Surveyor*)
Leland, O.M.	1904, 05, 06, 07, 08, 09, 10	IBC
Lewis, E.S.	1899	Harriman Expedition
Lord	1893	ABC
Mackie, F.H.	1908, 09, 10, 11	IBC
MacKinnon, J.S., Jr.	1908	
Maddren, A.G.	1913	USC&GS
Mansfield	1889	USC&GS
Martin, E.R.	1906, 07, 08	IBC
Martin, G.C.	1904, 05, 13	USGS
Martin, Lawrence	1904, 05, 09, 10, 11, 13	National Geographic Society and University of Wisconsin
Mather, K.F.	1923	(Alaska Peninsula)
McArthur, J.J.	1893, 94, 99, 1900, 01, 02, 04, 05, 06, 07, 08	DLS, ABC, IBC
McConnell, R.G.	1913	National Geographic Excursion
Mendenhall, W.C.	1898, 1902, 08	USGS (PWS)
Merriam, C. Hart	1899	Harriman Expedition
Mertie, J.B., Jr.	1911, 16, 17, 19	USGS
Moffit, F.H.	1904, 06, 19, 20, 22, 24, 25, 27, 28, 30	USGS
Moore	1913	Int'l Geological Congress
Morris & Joyce (?)	1887	
Morse, Fremont	1904, 05, 06, 07, 08, 10, 11, 13, 20	USC&GS; IBC
Mussell, H.S.	1904, 05, 06, 07, 08, 09, 10, 11, 12, 13, 14	IBC
Myers, C.L.	1906, 10	*Leslie's Weekly*, Colorado
Nelles, D.H.	1904, 05, 06, 07, 08, 09, 10, 12	IBC
Netland, L.	1904, 05, 06, 07, 08, 09	IBC
Ogilvie, William	1887, 88, 93, 94–95 (winter) 95, 96	DLS, ABC
Ogilvie, Noel J.	1905, 06, 09, 10, 12, 13, 14	DLS, IBC
O'Hara	1901, 02, 03, 11, 12	
Paige, Sidney	1903, 05	USGS (PWS)
Palache, Charles	1899	Harriman Expedition
Partridge, W.H.	ca. 1883, 87, 88	Commercial photographer, Portland, OR & Boston, MA
Patterson	1899	USC&GS

Photographer	Dates	Affiliation/Notes
Perkins, George W.	1909	
Pogue, J.E.	1913	USGS
Pond, Percy	1926	Winter & Pond Photo Studio, Juneau
Pounder, J.A.	1920	IBC
Pratt, J.F.	1893, 94	ABC
Pratt, Sherman	1931	With W.O. Field
Prince, F.W.	1907, 10	Pacific Coast Steamship Co.
Quillian, C.G.	1911	USC&GS
Radcliffe, H.	1905, 06, 07	IBC
Ratz, W.F.	1905, 06, 07, 08	IBC
Ray, L.L.	1931	USGS
Read, L.C.	1916	
Reed, John C.	1931	USGS
Reid, H.F.	1890, 92, 1931	Case Institute
Richardson, Wm. C.	1885	
Robertson, H.H.	1893, 94	ABC
Rockwell, Cleveland	1884 drawings	USC&GS
Russell, I.C.	1890, 91	USGS
St. Cyr, A.	1893, 94	DLS, ABC
Schrader, F.C.	1898	USGS (PWS)
Scidmore, Eliza R.	1886–87, 90	
Smith, P.S.	1911, 27, 28	USGS
Smith, W.R.		USGS
Snow, A.	1886/87	USN, USC&GS
Spencer, A.C.	1900	USGS (PWS)
Stanton, T.W.	1904	USGS
Stewart, B.O.	1925	(Glacier Bay)
Tabor, I.W.	1889	Probably commercial photographer
Talbot, A.C.	1893, 94	DLS, ABC
Tarr, Ralph S.	1905, 06, 08, 09, 10, 11, 13	USGS; Cornell University
Taylor, G. Morris	1916, 22, 28	
Thwaites, Frederick	1913	University of Wisconsin (with L. Martin)
Tittman, O.H.	1893, 1900, 04	ABC, IBC, USGS
Turrill & Miller	1905	San Francisco Photo
USC&GS	1889, 1911, 14, 16, 17, 20, 27, 29	
van Horn	1913	Int'l Geological Congress
Wales, Ralph	1919	
Wentworth, C.K.	1931	USGS
Wheeler, A.O.	1900, 03	DLS, Canada
White-Fraser, Geo.	1905, 06, 08	IBC
Whitney, P.C.	1907	USC&GS
Wilson, R.M.	1924	USGS
Wilson, Veazie	early 1890s	Juneau
Winter, Lloyd	1890s	Winter & Pond Photo Studio, Juneau
Worden, J.E.	1911	University of Wisconsin
Wright, C.W.	1903, 04, 05, 06, 31, 33	USGS
Wright, F.E.	1903, 04, 06	USGS
Wright, G.F.	1886	Oberlin College

The years listed above were taken from lists made by William O. Field of people in Alaska. The person could have been there in other years, but no information was found to verify that. Please note that there may not be pictures of glaciers taken in all the years given, especially for the IBC personnel. *Ed.*

APPENDIX E: EXPANDED NOTES

Chapter 3, note 7 (also Chapter 6, note 28; Chapter 8, note 22): History of the Alaska/Canada Boundary Survey Work

The settlement of the Alaska-Canada border is more recent than commonly believed and involved a tremendous amount of hard work and sacrifice on the part of the boundary survey parties. Between 1869 and 1920, about 150 Canadian and American government field parties of six to thirty men worked along the Alaska boundary. Today any part of the boundary can be reached in a few hours by helicopter and "it is easy to forget how remote it once was. For the surveyors, there were steamers along the coast and on the larger rivers, but beyond they used canoes, poling boats, pack-horses, or hand-drawn sleds and, at times, backpacked. Even the outboard motor was a luxury, first available in 1920, the final year of the survey."

The history of the boundary survey goes back to the Anglo-Russian Treaty of 1825, which fixed the boundary between British and Russian possessions on the North American continent. When the United States purchased the Russian possessions in 1867 and Canada took over the British possessions in 1870 and 1871, the boundary description given in the treaty continued unchanged. Later, the United States and Canada disagreed over the boundary description in what is now the Alaska panhandle as the language of the 1825 treaty was very indefinite in this area. Early in 1892 a proposal was accepted for a joint survey of the contested portion of the panhandle. This was to be done by the Alaska Boundary Commission (ABC) made up of both Americans and Canadians. A very poor field season in 1893 due to weather was followed by a very successful season in 1894, and a subsequent "clean-up" in 1895.

In 1899, following the surveys, a Joint High Commission was unable to settle the boundary dispute, and the differences culminated in the 1903 Alaska Boundary Tribunal. The six-member tribunal voted in favor of the United States, resulting in the *Atlas of Award*, published in 1904. "Then followed joint surveys by the United States and Canada (the International Boundary Commission, or IBC) from 1904 through 1920, to demarcate the boundary [as indicated in the *Atlas*]. The surveyors ignored the political differences and cooperated to complete a remarkable and nearly impossible task.

"By the close of the 1920 season, the field work had been completed and preparations for the final report and maps were underway. Thirteen maps were prepared. By early 1928 eleven of the thirteen maps had been completed and signed by the commissioners,

while the other two were held back for political rather than technical reasons" (See Green 1982).

For anyone interested in photographs of glaciers in Alaska, the boundary survey work is invaluable. The surveys were performed using a five-year-old photo-topographic method developed for mountain work. From a triangulated camera station, a series of photographs were taken covering a full 360°. In the 1890s, the U.S. field parties established camera stations high in the Coast Mountains that afforded clear views of the mountains and valleys farther inland. The result was great photographs of glaciers in the Coast Mountains dating from the 1890s. Copies of many of these photographs are included in Bill's collection now housed in the Archives of the Alaska and Polar Regions Department, Rasmuson Library, University of Alaska Fairbanks.

Chapter 4, note 1: Roadhouses

The Copper Center Roadhouse and Trading Post began as a temporary roadhouse established in 1898 by Andrew Holman near the confluence of the Klutina and Copper Rivers to provide shelter for prospectors who were coming into the area on their way to the goldfields. By the winter of 1899 Holman had built a substantial log hotel and a trading post. In 1901 the Hotel Holman and Trading Post became part of the assets of the new Copper River Mining, Trading and Development Company organized with Andrew Holman as president and Ringwald Blix as secretary-treasurer. Blix enlarged the hotel in 1906 and eventually bought the entire property. It was sold to Hans Ditman in 1918, and then to Mrs. Florence Barnes in 1923. She changed the name to Copper Center Roadhouse and Trading Post. The front half of the roadhouse was destroyed by fire in 1932 but a new two-story log building was erected on the same site. Later Mrs. Barnes also replaced the old part of the original building with a new addition that was used as the trading post. The new structure was known as the Copper Center Lodge and Trading Post and was an attractive building with diamond willow ornamenting the interior. Mrs. Barnes was famous for her hospitality. Mr. and Mrs. George Ashby purchased the property in May 1948 following Mrs. Barnes' death in February. They remodeled the trading post, making it into guestrooms. From that time it became known as the Copper Center Lodge. Mrs. Ashby has continued to operate the lodge since her husband's death in 1979.

A Gulkana Roadhouse was mentioned in a news item as early as the winter of 1904–1905. The roadhouse was strategically located: the Bear Creek Trail leading to Valdez Creek Mining district, the beginning of the Government Trail over the Alaska Range, and the junction of the Abercrombie Trail to Eagle all were located in this area. In 1907, Charles Levi Hoyt bought the roadhouse and together with Dolph Smith established the Gulkana Trading Post and Hotel. Hoyt became postmaster at Gulkana in 1909 and a lean-to was added on the west side of the hotel. Mrs. Elizabeth Griffith became the postmaster at Gulkana in 1916, and also became the new owner of the Gulkana Trading Post and Hotel. In 1931 a new two and one-half story log building was constructed and was named the Gulkana Lodge. After the war years, the O'Harra Bus Lines purchased the lodge to use as a relay point on its bus route and changed the name of the roadhouse to Santa Claus Lodge. It was destroyed by fire about 1948.

The Sourdough Roadhouse was established in 1903. When the new section of the Government Trail was constructed in 1906, it crossed the Alaska Range between the Copper River Valley and the Delta River. The Sourdough Roadhouse was the second roadhouse north from Gulkana and was listed as Pollard's on the Orr Stage Line distance table in 1906–1907. In 1908 Mrs. Nellie Yeager began to operate the roadhouse and it was known as the Sourdough Roadhouse and Trading Post. It was a single story log building with other outbuildings added through the years. The roadhouse was approved as a historical site by the state of Alaska in 1974, being the only establishment on the Richardson Highway built in the early days and still using the original building [see Phillips, Sr. (1984) for a preliminary report of the roadhouses established and operated on the government trail between Valdez and Fairbanks (filed in the Archives, Alaska and Polar Regions Department, Rasmuson Library, University of Alaska Fairbanks)].

Chapter 5, note 12: Bill's description of his assignment in Alaska during World War II

There is one more chapter in my filmmaking career: the war years. My entire three and one-quarter years in the Army during World War II were connected with filmmaking. I received a First Lieutenant's commission in the Signal Corp and my first job was at the Training Film Production Laboratory at Wright Field in Ohio. They were making training films for the Air Force, which was attached to the Army in those days. About half of the people at the lab were from Hollywood, and were used to making films on studio stages where you control the lighting and other things. The other half were from the East Coast, mostly newsreel people who also worked on such projects as documentary films. So this was a very interesting conglomeration, and in general it worked out. The first year at Wright Field was spent making various short training films on everything from how to adjust a carburetor to general flying tactics. I was primarily on the cameras, and other

people edited the film and put sound to it.

Late in December of 1943 I was sent to Alaska to work on a survival film. When I arrived in Fairbanks on the last day of December, two other lieutenants were already there. One was a writer and the other was a director from Hollywood, and they concocted the story that we filmed for the next three months. We were at Ladd Field, which is now Ft. Wainwright, in Fairbanks. It was the place to which fighters, transports and bombers were flown from the states up through Canada, for ferrying by Russian pilots to bases in the Soviet Union. Many of our pilots were from the South and were not used to operating in the cold and in areas covered with snow many months of the year. If they were forced down on their way to Alaska, they might be rescued in a few hours or a day but in the meantime, they'd frozen their hands and feet and might be out of service for a long time or permanently. So the plan was to make a film to show them how to handle themselves in such an emergency situation while waiting to be rescued. I was in charge of the camera crew and took most of the pictures myself. I had worked so much in remote places and under difficult conditions that I had more experience on how to make-do in a poor situation, and that was why I was sent up there. I found that the cameras had not been properly winterized as they were supposed to be, and we worked quite a long time to get them operating satisfactorily. It was so cold some days that if you just flicked the film while loading it into the camera, it would break. Also, it was January, which at best has only a few hours of sunlight low on the horizon in Fairbanks. We finished the film in early April. It was edited and released by that fall with the title *Land and Live in the Arctic*.

To make the film, we had to find a plane to use as the mock-up for our story, what you photograph around. By various filming tricks, we could take pictures in the air before the crash, and then the pilot in trouble and the plane coming down, but then most of the sequences took place around a plane on the ground that had crash landed and could not be flown away. Of course, we didn't want to use a brand new plane that was perfectly serviceable. But just about then, a young Russian pilot took off from Ladd Field for Nome, hit bad weather, turned around to came back and was blocked by the low, dense fog, which often happens in winter around Fairbanks. He missed the runway, crash-landed and then turned over a few miles from Ladd Field. We made do and changed the script to allow for the upside-down plane. There was a flurry of montage showing the crash landing in a blurred sort of way so we didn't have to show a lot of detail, and then the last shot of the crash was of the wheels still spinning. Then we showed what this pilot was supposed to do to protect himself, such as using oil from the plane for fuel, and taking parts of the fuselage off to flash a signal to a search plane flying overhead. A dog team then arrives for the rescue. There was a question in my mind as to whether the director from Hollywood, Jack Hively, had ever seen snow before, and in the process of filming the last sequence he wore boots with steel plates in them and as a consequence the soles of his feet froze. So he was in the hospital when we finished the final sequences of the film.

I remember a controversy over one point in the film. After the crash, do you first light a fire to warm yourself, or do you stop to change into dry socks, before the wet ones freeze? The so-called technical experts worked that out, but some disagreed. I don't remember the final decision but I do remember Stefansson telling me years later, "I'm not sure that you suggested the right procedure there at all." Of course, I didn't write the text, I just ran the movie camera. I don't think everyone ever would have agreed to the same opinion of what to do first!

Vilhjalmur Stefansson (1879–1962) was one of the last of the dog-team explorers of the Arctic. He spent more than half a century studying and exploring the Arctic. He wrote 24 books and at least 400 articles about the Far North and its people, including blond Eskimos. Stefansson began his northern explorations in 1904, and over the next fifteen years spent ten winters and thirteen summers in the Far North. In 1919 he retired from active exploration and devoted his energies to studying, writing and lecturing about the Arctic, and assembling the Stefansson Collection. The collection is now at Dartmouth College and is described by some authorities as the best and largest of its kind in the western world (New York Times, 27 August 1964). Ed.

Chapter 6, note 17: Wright's Report
Bill was always amused by the review of Wright's report:

The report contained a lot of very interesting information about the rate of sedimentation, the average rate the ice fields were lowering, and many pictures and maps. There were also a lot of comments in the margins, like it had been through the head USGS office for review. In one place was written, "Wright, we sent you up there to study geology, not to look at glaciers," and the report was never published. Wright also gave me many photographs he had taken on his trips and there were no captions on many of them but I knew where they were taken. Seeing those photos with no captions is what led me to start labeling my photos and mounting them in albums. It is no good to have photos around if they are not properly captioned because the material is next to useless. I can see the picture of a glacier and usually make a pretty good guess as to which one it is—each glacier has its own particular look. Austin Post could identify the glaciers too but to most people, one glacier looks like another.

Chapter 6, note 29: Sargent's maps
The following is a story Bill liked to relate concerning him, Sargent's map of Glacier Bay, and Mendenhall, the Director of the Geological Survey.

Sargent, who was the chief topographer of the Alaska branch of the Geological Survey had been in Glacier Bay in 1931 with John Reed, also with the Survey. I think they were doing ground control for the aerial pictures that had been taken by the navy in 1929. Sargent made a map of Glacier Bay based on his trip. It was drawn at only 1:20,000 on this thin paper and when I visited him one time at the Survey office in Washington, D.C. it was spread all over his room. I asked if I could make a tracing of some of the glacier fronts I was interested in, just for my own records, and he guessed it was all right. Later that day I was called into the director's office, Mendenhall. I walked in and he said, "Field, I understand you've been tracing our maps." I told him there were some ice fronts I wanted to get for my own information, and that I wasn't going to publish them or anything. He guessed it was okay, but he was pretty stern. I guess he was a gruff old man anyway. I don't think I got Sargent in trouble for letting me do this. I still have the original tracings.

Chapter 6, note 30: The naming of many of the glaciers in Johns Hopkins Inlet
The following is from Cooper (1937: 61–62).

Glacier Bay has heretofore been poorly supplied with place names. Accordingly, for convenience in description, Mr. Field and I have named those features to which reference is necessary and a few others of special interest.

Vicinity of Johns Hopkins Fiord (see Fig. 17): Johns Hopkins Fiord, formerly occupied by Johns Hopkins Glacier. Mt. Salisbury, for Rollin D. Salisbury, eminent student of Pleistocene and Recent glaciology. Mt. Abbe, for Cleveland Abbe, Jr., physiographer and climatologist, author of "The Climate of Alaska." Gilman Glacier, for Daniel C. Gilman, first president of the Johns Hopkins University. Clark Glacier, for William Bullock Clark, professor of geology at the Johns Hopkins University. Jones Glacier, for E. Lester Jones, member of the International Alaska-Canada Boundary Commission. Hoonah Glacier, for the Hoonah tribe of Indians, who live in the vicinity of Glacier Bay. Kashoto Glacier, for Kashoto, chief of the Hoonah tribe at the time of Muir's visit of 1879. Toyatte, Kadachan, Tyeen, John, and Charley Glaciers, for Indians associated with John Muir on his expeditions of 1879 and 1880.

Chapter 7, note 12: Richard C. Hubley
Richard C. Hubley (1926–1957) developed an early interest in the high mountains of western Washington and became fascinated by their natural characteristics, especially by the glaciers which adorn their slopes. He received his B.S. degree from the University of Washington in 1950 followed by his Master's degree in meteorology from the same institution in 1952. "By this time he had recognized the significant relationship that exists between the various forms of snow and ice and climatic environment. He thereupon extended his

graduate studies at the University of Washington in the field of micro-meteorology, and under the stimulating guidance of Phil E. Church and Franklin Badgley he was awarded the Ph.D. in 1957. An obituary by Walter A. Wood read:

> Although Dick's consuming interest in glaciers embraced all facets of their scientific study, his own research led him more and more towards an understanding of the exchange of mass and thermal energy that takes place between a glacier and its meteorological environment and from such studies to seek the relationships that exist between fluctuations of the glacier and those of the regional climate.
>
> In 1953 and 1954 Dick served as glaciologist of the Juneau Ice Field Research Project of the American Geographical Society and in the two succeeding years he received grants-in-aid from the Arctic Institute of North America to continue his pioneer work on the Blue Glacier of the Olympic Mountains....He was persuaded...to share in the planning and administrative phases of the U.S. program for the International Geophysical Year. As coordinator of the glaciological program of the USNC/IGY [U.S. National Committee for the IGY] he encouraged the concept of conducting concurrent glaciological studies throughout a wide range of latitude and environment and it was as a result of his urgings that the McCall Glacier became an object of IGY study under his scientific direction.
>
> North American science has been slow in building a coterie of scientists in the field of glaciology. Of those we have, Dick Hubley was one of the most distinguished.

Chapter 7, note 15: JIRP Reports issued by the Department of Exploration and Field Research of the American Geographical Society

Gilkey, Arthur K. (ed.) 1953. *Progress Report, Juneau Ice Field Research Project 1952 Season.*

Haley, Theodore R., Melvin G. Marcus, Maynard M. Miller, and Frederick A. Small. 1951. *Food Report, Juneau Ice Field Research Project, Alaska, June 1949–February 1951.* JIRP Rpt. No. 3.

Haley, Theodore R., Willard Nicholl, and Duncan L. McCollester. 1951. *Medical Reports, Juneau Ice Field Research Project, Alaska, June 1949–February 1951.* JIRP Rpt. No. 4.

LaChapelle, Edward R. 1954. *Snow Studies on the Juneau Ice Field.* JIRP Rpt. No. 9.

Miller, Maynard M. (ed.) 1949. *Progress Report of the Juneau Ice Field Research Project 1948.* JIRP Rpt. No. 1.

_____. 1950. *Preliminary Report of Field Operations of the Juneau Ice Field Research Project, Alaska, 1949 Field Season.*

_____. 1951. *Progress Report of Logistical Operations, Juneau Ice Field Research Project, Alaska, 1949, 1950, and 1951.* JIRP Rpt. No. 5

_____. (ed.) 1952. *Scientific Observations of the Juneau Ice Field Research Project, Alaska. 1949 Field Season.* JIRP Rpt. No. 6.

_____. 1953. *Juneau Ice Field Research Project, 1951 Winter Season.* JIRP Rpt. No. 8.

_____. 1954. *Juneau Ice Field Research Project, 1950 Summer Field Season.* JIRP Rpt. No. 7.

Nielsen, Lawrence E. (ed.) 1953. *Progress Report, Juneau Ice Field Research Project, 1953 Season.*

Publications directly or indirectly related to the project but issued elsewhere:

Bader, Henri. 1950. Introduction to Ice Petrofabrics. *Journal of Geology* 59: 519–536.

Field, William O., and Maynard Miller. 1950. The Juneau Ice Field Research Project. *Geographical Review* 40, no. 2: 179–190.

_____. 1951. Studies of the Taku Glacier, Alaska. *Journal of Geology* 59, no. 6: 622–623.

Field, William O. and Calvin J. Heusser. 1952. Glaciers: Historians of Climate. *Geographical Review* 42, no. 3: 337–345.

Gilkey, Arthur K. 1951. *Structural Observations on the Main Camp Nunatak, Juneau Ice Field, Alaska.* Unpub. M.S. thesis, Columbia University.

Heusser, Calvin J. 1952. Pollen Profiles from South-eastern Alaska. *Ecological Monographs* 22: 331–352.

_____. 1953. Radiocarbon Dating of the Thermal Maximum in Southeastern Alaska. *Ecology* 34: 637–640.

_____. 1954. Alpine Fir at Taku Glacier, Alaska, with Notes on its Postglacial Migration to the Territory. *Bulletin of the Torrey Botanical Club* 81, no. 1: 83–86.

_____. 1954. Additional Pollen Profiles from Southeastern Alaska. *American Journal of Science* 252: 106–119.

_____. 1954. Palynology of the Taku Glacier Snow Cover, Alaska, and its Significance in the Determination of Glacier Regimen. *American Journal of Science* 252: 291–308.

_____. 1954. Nunatak Flora of the Juneau Ice Field, Alaska. *Bulletin of the Torrey Botanical Club* 81, no. 3: 236–250.

Heusser, Calvin J., Robert L. Schuster, and Arthur K. Gilkey. 1954. Geobotanical Studies on the Taku Glacier Anomaly. *Geographical Review* 44, no. 2: 224–239.

Hubley, Richard C. 1954. Glaciers May Prove Helpful Key to World's Climatological Past. *Modern Precision* 14, no. 2: 8.

_____. 1955. Measurements of Diurnal Variations in Snow Albedo on Lemon Creek Glacier, Alaska. *Journal of Glaciology* 2: 560–563.

_____. 1956. An Analysis of Surface Energy During the Ablation Season on Lemon Creek Glacier, Alaska. *Transactions AGU* 38, no. 1 (Feb): 68–85. (Cont. #22, Dept of Meteorology and Climatology, Univ of Washington, Seattle.)

Lawrence, Donald B., and E.G. Lawrence. 1949. Some Glaciers of Southeastern Alaska. *Mazama* 31, no. 13: 24–30.

_____. 1950. Glacier Fluctuations for Six Centuries in Southeastern Alaska and its Relation to Solar Activity. *Geographical Review* 40: 191–223.

_____. 1951. Glacier Fluctuation in Northwestern North America Within the Past Six Centuries. *Union Géodésique et Géophysique Internationale, Assoc. Int. d'Hydrologie Scientifique, Assemblée Générale de Bruxelles*: 161–166.

Lawrence, Donald B., and John A. Elson. 1953. Periodicity of Deglaciation in North America since the late Wisconsin Maximum. *Geografiska Annaler* 35, no. 2: 83–104.

Leighton, F. Beach. 1951. Ogives of the East Twin Glacier, Alaska: Their Nature and Origin. *Journal of Geology* 59: 578–589.

Miller, Maynard M. 1949. 1948 Season of the Juneau Ice Field Research Project. *American Alpine Journal* 7: 185–191.

_____. 1951. Four Seasons on the Juneau Ice Field, Alaska. *The Explorers Journal*: 1–16.

_____. 1951. Winter Taku. *Appalachia*: 433–437.

_____. 1951. Adventure in Logistics. *The Mountaineer*: 15–19.

_____. 1951. Englacial Investigations Related to Core Drilling on the Upper Taku Glacier, Alaska. *Journal of Glaciology* 1, no. 10: 578–580.

_____. 1951. The Juneau Ice Field, Alaska, 1948–51. *American Alpine Journal* 8, no. 1: 113–118.

_____. 1951. Land of the Taku. *Canadian Alpine Journal*: 111–128.

_____. 1953. A Method for Bottom Sediment Sampling in Glacial Lakes. *Journal of Glaciology* 2, no. 14: 287–290.

Miller, Maynard M., and William O. Field. 1951. Exploring the Juneau Ice Cap. *Research Review*: 7–15.

Nichols, Robert L. and Maynard M. Miller. 1952. The Moreno Glacier. Lago Argentino, Patagonia. *Journal of Glaciology* 2, no. 11: 41–50.

Nielsen, Lawrence E. 1957. Preliminary Study on the Regimen and Movement of the Taku Glacier, Alaska. *Bulletin of the Geological Society of America* 68 (Feb): 171–180.

Poulter, Thomas C., C.F. Allen, and Stephen W. Miller. 1949. *Seismic Measurements on the Taku Glacier*. Stanford Research Institute.

Ward, Richard T. 1951. *Vegetation on Nunataks and Related Areas of the Juneau Ice Field, Alaska*. Unpub. M.S. thesis, University of Minnesota.

Chapter 8, note 13: The letter describing first-hand the giant wave in Lituya Bay

"Dear Bernice and Bob,

There has been very little chance to write this summer, so am taking advantage of this day in the harbor. . . . No doubt you have read about the terrible earthquake and tidal wave that we experienced up here. . . . We are thankful to be alive. I might not be writing this letter today if we had gone ahead with our original plans for that fatal day.

"We were all alone in Lituya Bay on July 9th but left there to try the fishing outside of the harbor. Finding no fish there, we decided to go off shore, so ran for three hours out from Icy Point. We had intended to go back in that night, and go prospecting the next day—the kings [salmon] seemed to have completely left the country up there and the big fleet from the States had gone. . . . It must have been God's will that the kings disappear at this time or the whole fleet would have been wiped out, and that would have included us. For three weeks there were about sixty boats lying in Lituya Bay while on the day of the quake there were only three. One managed to ride it out behind a point, one was carried over La Chausee Spit and swamped outside—but the couple got in their skiff and were picked up by the only troller passing at the time, the third couple lost their lives as their boat was swept sideways over the Spit. We always laid behind that Spit and were there only a few hours before the quake. We had planned to come back in time to pick our strawberries that night. I had remarked to John how lovely and peaceful it was there. Now nothing remains but Desolation and Chaos. Beautiful Lituya is no more. Every living thing, including trees, was swept clean—down to the bedrock.

"I will review events briefly as they happened: We had our gear in the water off shore and were 'in the fish.' We were both in the wheel house waiting for the lines to load up again and remarking that it was ominously calm and humid that night. At 10:17 P.M. we were lifted up like match-sticks and shaken so violently that we thought that the boat would to go pieces. We could see the shock wave coming over the water towards us. We were all alone out here and we wondered how much our boat could stand. When it struck it knocked us off our feet and continued to shake till we thought for sure our boat would fall apart. We hung on for all we were worth and offered a silent prayer. Then we turned on the radio to see if we could learn anything about what had happened and what was going on at the time. Excited voices filled the air, and cries of 'May Day, May Day' filled the air. Some voices were hysterical and some calm. Finally others were quiet long enough for this message to come through, 'My God, it looks like the end of the world in here. I am in Lituya Bay and a tidal wave 100 feet high is sweeping the Bay. I lost my anchor when we were lifted up and do not know whether we will make it or not. Tell my wife in the *Pelican*. This is the boat *Edrie*' [see *Tacoma*

News Tribune, March 21, 1982: 12–15 for the story of Howard Ulrich, the skipper of the *Edrie*]. Then there was silence awhile and then came, 'Thank God, I made it. I rode it out in deep water, but now I must get out of here as there are huge logs and trees drifting past me. If there are any boats behind the Spit, they have had it.' Our ship to ship frequency had gone out so we could not talk to other ships and the next thing we heard was: 'The *Cameo* was in there, several people saw them go in. No doubt they are done for now.' We stood by the radio listening to our demise and unable to do anything about it or tell people that we were still alive.

"As things quieted a bit we tuned in our ship-to-shore and finally got the *Pelican* and informed them we were still alive.... Juneau was badly shaken too but was further [*sic*] from the center of the quake. Pelican and Elfin Cove were badly shaken too and rocks came rolling down the hillside. All of Southeastern Alaska felt it and there were many slides down from the hills.... We heard the horrible news that two of our friends didn't make it, and two others were miraculously saved. Both those boats had been anchored behind the Spit and were swept completely over it, even above the tree tops. The Spit is about 100 feet wide and 200 feet high and was covered with big boulders and tall trees. Both of these vessels were large ocean-going trollers, forty-seven-feet long. [The couple that was lost] was in bed when they heard the roar of the glacier and the mountain breaking. They started their engine and headed towards the harbor entrance but met the tidal wave that carried them across the Spit. It was loaded with trees and debris. They went across the spit sideways and were swamped. The next morning all that remained of them was some oil slick and bubbles outside the harbor.

"The other couple had such a remarkable deliverance that it is hard to believe. Somehow their boat rode the wave over the spit to the outside, then it broke in two. A log came through the pilot house and struck him in the chest, yet they managed to get their skiff loose. As their boat sank they got into the skiff, which turned over and left them clinging to it. A lone troller [trawler] was passing by and when the skipper saw a light he went to investigate and found them. There had been no light that anyone else saw—yet it led him to them. That is the strange part. They soon recovered, but lost everything.... Believe me that couple are going to do a lot of serious thinking and wondering how they ever came out of this. It was not just an accident that one troller happened to be in that vicinity that night. They would not have survived long in that icy water....

"A bush pilot friend flew over the area the next day and reported that a whole mountain had fallen into the water along with the glacier, and he believed that was what caused the great tidal wave. We are also inclined to believe there is a fault that lies under the Bay, and this has happened before, once in 1853 when an entire Indian village was swept away, and once in 1936 when it did some damage but nothing like this.... Three other people were picking berries on an island near Yakutat and one-half of the island lifted up about twenty feet, then dropped down and disappeared, taking them with it. No trace of them has ever been found. Our [pilot] friend also reported the wave had swept 1800 feet up one side of Lituya Bay and 900 feet up the other side, carrying away every tree and sweeping it bare to the bedrock. The trees which litter the Gulf and cross the Sound for miles are stripped clean of all bark; some have been up-rooted, others were ground to match size and kindling wood. All are clean and white...just like they came from a peeler, not a shred of bark or foliage left on any of [them]....

"One thing that I forgot to mention is that as we looked towards shore we saw what appeared to be a whole mountain fall into the sea, and as other great pieces of the mountain and ice rolled down dust rose thousands of feet and almost obliterated the shoreline.... We stayed all night by the radio and shock after shock occurred—not as bad as the first but bad enough.

"Gone are the beautiful Sabbaths that we spent at Lituya Bay. It was a place of peace and beauty—now it is a scene of desolation. Hundreds perhaps thousands of seal, bear and other wildlife must have perished in the destruction caused by the quake. Gone is the great bird rookery there and gone is our strawberry patch and the home we had there... it will never again be a peaceful refuge, but a dangerous haven in time of storm."

BIBLIOGRAPHY

A Voyage Round the World. Performed in the years 1785, 1786, 1787, and 1788 by the Boussole *and* Astrolabe *under the command of J.F.G. de la Pérouse.* 1799. In 2 volumes. Translated from the French, printed by A. Hamilton for G.G. and J. Robinson, Paternoster-Row; J. Edwards, Pall-Mall; and T. Payne, Mews-Gate, Castle-Street.

Allen, W.E.D. 1927. New Political Boundaries in the Caucasus. *Geographical Journal* 69, no. 5 (May): 430–442, 1 map.

_____. 1932. *A History of the Georgian People from the Beginning down to the Russian Conquest in the Nineteenth Century.* Kegan Paul, Trench, Trubner & Co., Ltd., London.

Allen, W.E.D., and Paul Muratoff. 1953. *Caucasian Battle Fields: A History of the Wars on the Torco-Caucasian Border, 1828–1921.* Cambridge University Press, Cambridge.

American Geographical Society. 1960. *Nine Glacier Maps, Northwestern North America.* Special Publication no. 34.

Andrews, Clarence L. 1903. Muir Glacier. *National Geographic Magazine* 14, no.12: 441–445.

Appalachia. 1902. Vol. 10, no. 1: 28–43 and frontispiece.

Appalachia. 1938. Vol. 22, no. 86 (Dec): 287 and photo facing 285.

Arctic and Alpine Research. 1983. Vol. 15, no. 4: 425–554.

Atwood, Evangeline, and Robert N. DeArmond. 1977. *Who's Who in Alaskan Politics.* Alaska Historical Commission, Anchorage.

Baddeley, John Frederick. 1908. *The Russian Conquest of the Caucasus.* Longmans. (Reprinted by Russell & Russell, 1969.)

_____. 1940. *The Rugged Flanks of the Caucasus.* 2 vols. Oxford University Press, London.

Baldwin, S. Prentiss. 1893. Recent changes in the Muir Glacier. *American Geologist* 11 (June): 366–375.

Block, Maxine, ed. 1940. *Current Biography: Who's News and Why.* H.W. Wilson Co, New York.

Bohn, Dave, ed. 1958. The Juneau Ice Field Research Project. *Mazama* 40, no. 13 (December).

Bohn, Dave (text and photographs), and David Brower (ed.). 1967. *Glacier Bay: The Land and the Silence.* Alaska National Parks and Monuments Association, Anchorage.

Brown, William L. 1929. The Beach Anglo-Egyptian Sudan Expedition. In *Explorations and Field-work of the Smithsonian Institution in 1928,* 63–70. Smithsonian Institution Publication 3011, Washington, D.C.

Burroughs, John et al. (eds.) 1986. *Alaska: The Harriman Expedition, 1899.* Dover Publications, Inc., New York.

Carpé, Allen. 1931. The Conquest of Mt. Fairweather. *American Alpine Journal* 43 (November): 221–231, 3 plates.

_____. 1933. The Mt. Logan Adventure. *American Alpine Journal.*

Cashen, William R. 1971. *Farthest North College President.* University of Alaska Press, Fairbanks.

Chorlton, Windsor, and the Editors of Time-Life Books. 1983. *Planet Earth: Ice Ages.* Time-Life Books, Alexandria, VA.

Clayton, Lee, William O. Field, and Samuel J. Tuthill. 1966. Recent Fluctuations of the Sherman and Sheridan Glaciers, South-Central Alaska, (abstract). *Transactions of the American Geophysical Union* 47: 81–82.

Coch, Nicholas K., and Allan Ludman. 1992. *Physical Geology.* Macmillan Publishing Company, New York.

Cooper, William S. 1923. The Recent Ecological History of Glacier Bay, Alaska: I. The Interglacial Forests of Glacier Bay. *Ecology* 4, no. 2 (April): 93–128.

_____. 1931. A Third Expedition to Glacier Bay. *Ecology* 12, no. 1 (January): 61–95.

_____. 1937. The Problem of Glacier Bay, Alaska: A Study of Glacier Variations. *Geographical Review* 27, no. 1: 37–62.

_____. 1942. Vegetation of the Prince William Sound Region, Alaska, with a Brief Excursion into Post-Pleistocene Climatic History. *Ecological Monographs* 12, no. 1 (January): 1–22.

_____. 1956. *A Contribution to the History of the Glacier Bay National Monument.* Department of Botany, University of Minnesota (March).

Crary, Albert P., William O. Field, and Mark F. Meier. 1962. The United States Glaciological Researches During the International Geophysical Year. *Journal of Glaciology* 4, no. 31 (March).

Cushing, Henry P. 1891. Notes on the Muir Glacier Region and its Geology. *American Geologist* 8, no. 4 (October): 207–230.

Davidson, George. 1904. The Glaciers of Alaska that are Shown on Russian Charts or Mentioned in Older Narratives. *Transactions and Proceedings of the Geo-Graphical Society of the Pacific* 3, series 2 (June): 2–98.

Dunn, J.M. 1949. *The Russian Revolution.* Lucent.

Ferris, Benjamin G. 1947. Mount St. Elias. *Harvard Mountaineering* 8 (May): 7–21.

Field, Alice Withrow. 1932. *Protection of Women and Children in Soviet Russia.* Dutton.

Field, Frederick Vanderbilt. 1983. *From Right to Left: An Autobiography.* L. Hill.

Field, William O., Jr. 1926. The Fairweather Range: Mountaineering and Glacier Studies. *Appalachia,* 16 (December): 460–472.

————. 1927. In Search of Mount Clearwater. *Harvard Mountaineering* 1, no. 1 (June): 5–11.

————. 1930. Travels and Mountaineering in the Caucasus. *American Alpine Journal* 1, no. 2: 167–173.

————. 1932. The Glaciers of the Northern Part of Prince William Sound, Alaska. *Geographical Review* 22, no. 3 (July): 361–388.

————. 1932. The Mountains and Glaciers of Prince William Sound, Alaska. *American Alpine Journal* 1, no. 4: 445–458.

————. 1937. Observations on Alaskan Coastal Glaciers in 1935. *Geographical Review* 27, no. 1 (January): 63–81.

————. 1947. Glacier Recession in Muir Inlet, Glacier Bay, Alaska. *Geographical Review* 37, no. 3: 369–399.

————. 1949. Glacier Observations in the Canadian Rockies. *Canadian Alpine Journal:* 99–114.

Field, William O. 1958. "Flight to the South Pole." *Quad* 12, no. 2 (May). Riverdale County School, Pinedale-on-Hudson, New York.

————. 1965. Avalanches caused by the Alaska Earthquake of March 1964. International Symposium on Scientific Aspects of Snow and Ice Avalanches. *International Association of Scientific Hydrology Publication* 69: 326–331. International Association of Scientific Hydrology, Gentbrugge, Belgium.

————. 1968. The Effect of Previous Earthquakes on Glaciers. In *The Great Alaska Earthquake of 1964,* vol. 3, *Hydrology,* pp. 252-265. National Academy of Sciences Publication 1603, Washington, D.C.

————. ed. 1975. *Mountain Glaciers of the Northern Hemisphere.* 3 vols. Cold Regions Research and Engineering Laboratory, U.S. Army Corps of Engineers, Hanover, NH.

Field, William O., Jr., and Calvin J. Heusser. 1954. Glacier and Botanical Studies in the Canadian Rockies, 1953. *Canadian Alpine Journal* 37: 128–140.

Field, William O., Jr., et al. 1958. *Geographic Study of Mountain Glaciation in the Northern Hemisphere.* American Geographical Society, N.Y., for the U.S. Army Quartermaster Research and Development Command, 10 parts in 3 vols.

————. 1958. *Atlas of Moutain Glaciers in the Northern Hemisphere.* American Geographical Society, N.Y., for the U.S. Army Quartermaster Research and Development Command.

Fritiof, Fryxell. 1956. Memorial to François Emile Matthes. In *Proceedings of the Geological Society of America, Annual Report for 1955* (July): 153–168.

Freeman, Lewis R. 1925. *On the Roof of the Rockies.* Dodd, Mead and Company, New York.

Freshfield, Douglas W. 1896. *The Exploration of the Caucasus.* 2 vols. Edward Arnold, London.

Geographical Review. 1955. Vol. 45, no. 4 (October).

Geological Society of America Memorials. 1973. Vol 1: 97–104.

Geological Society of America Memorials. 1980. Vol 10: 1–8.

Geological Society of America Memorials. 1982. Vol 12.

Geological Society of America Memorials. 1984. Vol 14: 1–8.

Geological Society of America. 1980. *Memorial to William Skinner Cooper.* March.

Gilbert, Grove Karl. 1904. *Glaciers and Glaciation.* Vol. 3 of *Harriman Alaska Expedition,* C. Hart Merriam, ed. Doubleday, Page and Company, New York.

Green, Lewis. 1982. *The Boundary Hunters.* University of British Columbia Press, Vancouver.

Grant, Ulysses S., and D.F. Higgins. 1913. *Coastal Glaciers of Prince William Sound and Kenai Peninsula, Alaska.* U.S. Geological Survey Bulletin 526.

Harris, Lement. *My Tale of Two Worlds.* 1986. International Publishers, New York.

Heusser, Calvin J. 1951. *Glacier Fluctuations in the Canadian Rockies.* Publication no. 39 de l'Association Internationale d'Hydrologie Scientifique (Assemblée Générale de Rome, tome IV): 493–497.

ICE (News Bulletin of the International Glaciological Society). 1972. No. 38: 16.

Imbrie, John, and Katherine P. Imbrie. 1979. *Ice Ages.* Enslow Publishers, Hillside, NJ.

Janson, Lone E. 1979. *The Copper Spike.* Robert N. DeArmond, ed. Alaska Northwest Publishing, Anchorage.

Keen (Handy), Dora. 1910. A Woman's Ascent of the Matterhorn. *Outlook*.

_____. 1911. A Woman's Climbs in the High Alps. *National Geographic Magazine* (July).

_____. 1912. Arctic Mountaineering by a Woman. *Scribner's* (May).

_____. 1912. First Expedition to Mt. Blackburn, Alaska. *Appalachia* 12, no. 4 (April).

_____. 1915. Exploring the Harvard Glacier, 1914. *Harper's Magazine* 132 (Dec): 113–125.

_____. 1915. Studying the Alaskan Glaciers. *Philadelphia Geographical Society Bulletin* 13, no. 2: 25–32.

_____. 1915. First Exploration of the Harvard Glacier, Alaska. *Bulletin of the American Geographical Society* 47, no. 2 (February): 117–119.

Klotz, Otto. 1907. Recession of Alaskan Glaciers. *Geographical Journal* 30: 419–421.

Ladd, William S. 1932. The Fairweather Climb. *American Alpine Journal* 1, no. 4: 428–443.

Lampart, H.F.J. 1926. Conquest of Mt. Logan, 1925. *Geographical Journal* 68 (July): 1–26, 8 plates.

Lamplugh, George W. 1886. Notes on the Muir Glacier of Alaska. *Nature* 28 (January): 299–301.

Marangunic, Cedomir, and Colin Bull. 1966. The Earthquake-Induced Slide of the Sherman Glacier, S. Alaska, and its Glaciological Effects, (abstract). *Transactions of the American Geophysical Union* 47: 81.

Marangunic, Cedomir, and Colin Bull. 1968. The Landslide on the Sherman Glacier. In *The Great Alaska Earthquake of 1964*, vol. 3, *Hydrology*, pp. 383–394. National Academy of Sciences Publication 1603, Washington, D.C.

Marshack, Alexander. 1958. *The World in Space: The Story of the International Geophysical Year.* Thomas Nelson & Sons, New York.

Martin, Lawrence. 1913. Alaskan Glaciers in Relation to Life. *Bulletin of the American Geographical Society* 45, no. 11: 801–818.

Martin, Lawrence. 1913. "Glaciers and International Boundaries." *Scientific American Supplement* 76, no. 1965 (August 30): 135–138.

Matthes, François E. 1960. *Geologic History of the Yosemite Valley.* U.S. Geological Survey Professional Paper 160.

McManis, Douglas R. 1990. The Editorial Legacy of Gladys M. Wrigley. *Geographical Review* 80, no. 2 (April) 169–181.

Meek, Victor. 1948. Glacier Observations on the Canadian Cordillera. *Canadian Geographical Journal* 37, no. 5: 190–209.

Meier, Mark F., 1984. Contribution of Small Glaciers to Global Sea Level. *Science* 226: 1418.

Memorial to Harry Fielding Reid. 1944. *Proceedings of the Geological Society of America*: 295–298.

Mendenhall, Walter C. Memorial of Grove Karl Gilbert. *Proceedings of the Boston Meeting of the Geological Society of America*: 26–44.

Mercanton, P. Louis. 1916. *Measurements of the Rhône Glacier 1874 to 1915.* The Commission on Glaciers of the Swiss Society (Academy) of Natural Sciences.

Merriam, C. Hart, ed. 1901–1914. *Harriman Alaska Expedition.* Vol. 1, Alaska: Narrative, Glaciers, Natives; vol. 2 (general description of the expedition's findings); vol. 3, Glaciers and Glaciation; vol. 4, Geology and Paleontology; vol. 5, Cryptogamic Botany; vol. 8, Insects, pt. 1; vol. 9, Insects, pt. 2; vol. 10, Crustaceans; vol. 11, Nemertians and Bryozoans; vol. 12, Enchytraeids and Tribicolous Annelids; vol. 13, The Shallow-water Starfishes of the North Pacific Coast From the Arctic Ocean to California, 2 bks. Doubleday, Page and Company, New York, and Smithsonian Institution, Washington, D.C.

Miller, Don J. 1960. *Giant Waves in Lituya Bay, Alaska.* U.S. Geological Survey Professional Paper 354-C.

Miller, Maynard M. 1947. Yahtsetesha. *American Alpine Journal* 4, no. 3: 257–268.

Miller, Orlando W. 1973. *The Frontier in Alaska and the Matanuska Colony.* Yale University Press, New Haven.

Mitchell, B.W. 1924. *Trail Life in the Canadian Rockies.* MacMillan Co., New York.

Molenaar, Dee. 1947. St. Elias: The First American Ascent. *Sierra Club Bulletin* 32, no. 5 (May): 62–70.

Moore, Terris. 1931. Mt. Fairweather is Conquered at Last. *The Sportsman* 10 (October): 48.

Morse, Fremont. 1908. The Recession of the Glaciers of Glacier Bay, Alaska. *National Geographic Magazine* 19: 76–78.

Mummery, A.F. 1895. *My Climbs in the Alps and Caucasus.* Charles Scribner's Sons, New York.

New York Post, 8 April 1935.

New York Times, 7 January 1949.

New York Times, 11 May 1949, obituary.

New York Times, 27 August 1964.

Orth, Donald J. 1971. *Dictionary of Alaska Place Names.* U.S. Geological Survey Professional Paper 567, U.S. Government Printing Office, Washington, D.C.

Patton, Brian, ed. 1984. *Tales from the Canadian Rockies*. Hurtig Publishers Ltd., Alberta.

Phillips, Walter T., Sr. 1984. *Roadhouses of the Richardson Highway, 1898–1923*. A preliminary report of the roadhouses established and operated on the government trail between Valdez and Fairbanks. On file in the Archives, Alaska and Polar Regions Department, Rasmuson Library, University of Alaska Fairbanks.

Plafker, George. 1968. Source Areas of the Shattered Peak and Pyramid Peak Landslides at Sherman Glacier. In *The Great Alaska Earthquake of 1964*, vol. 3, *Hydrology*, pp. 374–382. National Academy of Sciences Publication 1603, Washington, D.C.

Pole, Graeme. 1991. *The Canadian Rockies: A History in Photographs*. Altitude Publishing, Canadian Rockies/Vancouver.

Reid, Harry Fielding. 1892. Report of an Expedition to Muir Glacier, Alaska, with Determinations of Latitude and the Magnetic Elements at Camp Muir, Glacier Bay. *Report of the Superintendent of the U.S. Coast and Geodetic Survey, Part 2*. U.S. Government Printing Office, Washington, D.C.

_____. 1892. Studies of Muir Glacier, Alaska. *National Geographic Magazine* 4 (March): 19–84.

_____. 1896. Glacier Bay and its Glaciers. In C. D. Walcott, ed., *U.S. Geological Survey 16th Annual Report, 1894–95*, part 1, 421–461. Washington, D.C.

Russell, Israel C. 1897. *Glaciers of North America*. Ginn & Company.

Schäffer, Mary. 1908. *Canadian Alpine Journal* 1, no. 2: 288.

_____. 1911. *Old Indian Trails*. New York.

Shepard, Francis P., and Harold R. Wanless. 1971. *Our Changing Coastlines*. McGraw-Hill Book Co., New York.

Shreve, Ron L. 1966. Sherman Landslide, Alaska. *Science* 154: 1639–1401.

Spectrum. July–August 1978, SRI International.

Stephen, Sir Leslie, and Sir Sidney Lee, eds. *Dictionary of National Biography from Earliest Times to 1900*. Vol. 7. Oxford University Press, London.

Stutfield, H.E.M., and J. Norman Collie, F.R.S. 1903. *Climbs and Exploration in the Canadian Rockies*. Longmans, Green and Co., London.

Tarr, Ralph S., and Lawrence Martin. 1907. Position of Hubbard Glacier Front in 1792 and 1794. *Bulletin of the American Geographical Society* 39 (March): 1–8.

_____. 1912. *The Earthquakes at Yakutat Bay, Alaska, in September 1899*. U.S. Geological Survey Professional Paper 69.

_____. 1914. *Alaskan Glacier Studies*. National Geographic Society, Washington, D.C.

_____. 1914. *Alaskan Glacier Studies of the National Geographic Society in The Yakutat Bay, Prince William Sound and Lower Copper River Regions*. National Geographic Society, Washington, D.C.

The Expedition of the National Geographic Society and the U.S. Geological Survey (1890). 1891. *Century Magazine* 41 (April): 872.

Thorington, J. Monroe, 1925. *The Glittering Mountains of Canada: A Record of Exploration and Pioneer Ascents in the Canadian Rockies, 1914–1924*. John W. Lea, Philadelphia.

Tuthill, Samuel J. 1966. Sherman Glacier, Paleoecologic Laboratory (Abstract). *Transactions of the American Geophysical Union* 47: 82.

Tuthill, Samuel J., William O. Field, and Lee Clayton. 1968. Postearthquake Studies at Sherman and Sheridan Glaciers. In *The Great Alaska Earthquake of 1964*, vol. 3, *Hydrology*, pp. 318–328. National Academy of Sciences Publication 1603, Washington, D.C.

Vancouver, George. 1798. Glacier Bay. In *A Voyage of Discovery to the North Pacific Ocean and Around the World*, vol. 3, 405–452. Robinson and Edwards, London.

von Dechy, Moriz. 1905. *Kaukasus: Reisen und Forschungen im Kaukasischen Hochgebirge*. 3 vols. Dietrich Reimer, Berlin.

Wentworth, Chester K., and Louis L. Ray. 1936. Studies of Certain Alaskan Glaciers in 1931. *Bulletin of the Geological Society of America* 47: 879–934.

Wheeler, Arthur O. 1931. *Canadian Alpine Journal* 20: 120–142.

Wood, Walter. 1942. The Parachuting of Expedition Supplies: An Experiment by the Wood Yukon Expedition of 1941. *Geographical Review* 32, no. 1: 36.

Wright, George F. 1887. The Muir Glacier. *American Journal of Science*, third series, vol 33, no. 193: 1–18.

Wright, John K. 1952. *Geography in the Making: The American Geographical Society, 1852–1952*. American Geographical Society, New York.

_____. 1952. The American Geographical Society: 1852–1952. *Scientific Monthly* 74, no. 3 (March): 121–132.

INDEX

WOF refers to William Osgood Field.
Bold page numbers refer to photographs.